QUANTUM CHEMISTRY

Quantum Chemistry

The development of *ab initio* methods in molecular electronic structure theory

HENRY F. SCHAEFER III

Department of Chemistry
University of California
Berkeley California

CLARENDON PRESS · OXFORD
1984

Oxford University Press, Walton Street, Oxford OX2 6DP

London Glasgow New York Toronto
Delhi Bombay Calcutta Madras Karachi
Kuala Lumpur Singapore Hong Kong Tokyo
Nairobi Dar es Salaam Cape Town
Melbourne Auckland

and associated companies in
Beirut Berlin Ibadan Mexico City Nicosia

Oxford is a trade mark of Oxford University Press

Published in the United States
by Oxford University Press, New York

ISBN 0 19 855183 5

British Library Cataloguing in Publication Data

Schaefer, Henry F.
Quantum chemistry.
1. Quantum chemistry
I. Title
541.2'8 QD462

ISBN 0-19-855183-5

Library of Congress Cataloging in Publication Data

Schaefer, Henry F.
Quantum chemistry.

Bibliography: p.
1. Quantum chemistry. 2. Electronic structure.
I. Title.
QD462.S32 1984 541.2'8 84-925
ISBN 0-19-855183-5

Set by Joshua Associates, Oxford
Printed in Great Britain
by J. W. Arrowsmith Limited, Bristol

Preface

More than eleven years ago, the author completed the monograph, *The electronic structure of atoms and molecules: a survey of rigorous quantum mechanical results*. At the time of publication (1972), most practicing chemists were not aware that the *ab initio* methods of molecular electronic structure theory were capable of providing reliable predictions of quantities of chemical interest. A number of my colleagues at Berkeley expressed surprise at the time that variational quantum mechanical methods could be applied to systems as large as XeF_6, azulene, and the guanine-cytosine base pair. Similar astonishment greeted the notion that for some of the simplest polyatomic molecules (e.g. methylene) theory could successfully challenge the most sophisticated experimental spectroscopic studies.

During the past decade the situation described above has almost entirely reversed itself. In what has been described as the beginning of the 'Third Age of Quantum Chemistry' [W. G. Richards, Nature **278**, 507 (1979)], there are now literally dozens of cases in which electronic structure theorists have successfully contested the laboratory conclusions of distinguished experimentalists. The specific case Richards describes in the above-cited article is one in which theory and experiment worked closely together to unravel the spectroscopy of triplet acetylene in a manner which would not have been possible by either party working in isolation.

Another sign of the minor chemical revolution that *ab initio* methods are producing is the fact that a number of highly regarded experimental chemists have more or less abandoned their laboratory research programs in favor of quantum mechanical studies. In most cases this transformation from experiment to theory has been quite successful, since experimentalists appreciate the subtleties of their own laboratory equipment and tend to transfer this healthy caution to the theoretical methods they adopt when going into the *ab initio* field on a full-time basis.

However, the majority of users of electronic structure theory are no longer professional theoreticians, but rather experimentalists who simply do not have the time to take off a year or two to become experts in the field. And this is precisely the group whose numbers will grow at by far the most rapid rate during the next decade. This phenomenon must be considered simultaneously the greatest triumph and the greatest threat to the integrity of molecular quantum mechanics. The use of the former description should be obvious, since the goal of most electronic structure theorists since 1930 has been the development of methods readily applicable to the broadest range of chemical problems.

Nevertheless, one must concede that there are great perils associated with the application of *ab initio* methods to chemical problems. When an experimentalist turns in his latest new compound to the departmental n.m.r. facility, at least he is reasonably certain that the output will be an n.m.r. spectrum. He may have to be careful that the n.m.r. spectrum refers to a single compound rather than a mixture, but this is an uncertainty that chemists have learned to deal with. In striking contrast, it is very easy to submit a deck of computer cards to a standard quantum mechanical program such as GAUSSIAN 70 and receive as output totally meaningless results. In fact, it is probably not an exaggeration to state that there are literally hundreds of such error-ridden calculations which have actually been published in the chemical literature. A cynical view of this situation is provided by the statement that 'running a few molecular orbital calculations doesn't make one an electronic structure theorist any more than sleeping in one's garage makes him/her an automobile' (anonymous, 1981).

It would seem to the present writer that the best deterrent to the sort of disasters described above is a vigorous program of theoretical education. The present volume resulted from a request from my graduate students and postdoctorals at Berkeley for an informal lecture on the history of *ab initio* methods in theoretical chemistry. I initially thought this could be accomplished in one hour, but found that only the period through 1965 was surveyed in that short period. At the conclusion of this first talk, there were several requests for references to the original papers to which I had referred. In retrospect, it became clear that such a chronological approach, based on an (obviously incomplete) sequence of landmark papers, might be a helpful approach to discussions of the origins of *ab initio* methods in molecular quantum mechanics.

The hope is that this collection will be of interest to the broad spectrum of experimental chemists who occasionally use electronic structure theory as an aid to the interpretation of their laboratory findings. In this regard it should be possible to scan the entire book in a single day. It should go without saying that the hope of this work is that the reader will go back and study in their entirety some of the original papers cited here. In fact the original idea was to publish all of the selected papers in reprint form. However as the (incomplete) list of landmark papers approached 150, it became obvious that such a procedure would result in an excessively lengthy and expensive volume. Therefore it was decided to include only the first page (typically spanning abstract plus introduction) of each paper when the manuscript was distributed to members of my research group.

In addition to the abstract, a brief commentary was given on each paper, giving some indication of its broad significance, with a few references to important related research. Ultimately, space considerations mandated that only the comments be reproduced in the present volume. A perusal of these comments will demonstrate to the reader that the primary focus is on methodology rather than the examination of particular chemical problems. The papers chosen here either present new and important methods or illustrate how well (or poorly) existing methods do in predicting various chemical phenomena. Note also that certain papers are so well known (for example, those by Hartree, Slater, Fock, Mulliken, Heitler, London, Brillouin, and Koopmans) and appreciated that inclusion here would be superfluous.

Finally, I would request a spirit of tolerance from my colleagues, the professional molecular electronic structure theorists. The distinction between a landmark or classic paper and one that is 'merely' very important represents a treacherous and very grey area. Thus the task of choosing 150 papers from more than one thousand which have been of significant help in my own research is a very difficult one. I have requested the advice of a number of colleagues, but choose only to thank them anonymously here, as I take responsibility for the highly subjective choice of papers presented here. In the event that a second edition of this volume is called for, suggestions would be sincerely appreciated.

I am grateful to Mrs Carol Hacker for typing the manuscript and to Dr Douglas J. Fox and Professor Russell M. Pitzer for their careful reading of the manuscript. The comments of several other members of my research group are also appreciated.

Berkeley H.F.S.
June 1983

In memory of Pierre Edward Schaefer
(June 20–December 9, 1979)

Let the little children come to Me,
and do not forbid them; for of such
is the kingdom of God. (Luke 18:16)

An incomplete list of landmark papers in *ab initio* molecular electronic structure methods

The order of entry is first by year, then by journal (with journals listed in alphabetical order), and finally within each journal-year by page number of appearance. Note that this (arbitrary) ranking can (and occasionally does) result in a failure to maintain accepted scientific priority.

1928

1. E. A. Hylleraas, Über den Grundzustand des Heliumatoms, *Z. Physik*. **48**, 469.

1933

2 H. M. James and A. S. Coolidge, The ground state of the hydrogen molecule, *J. Chem. Phys.* **1**, 825.

1934

3 C. Møller and M. S. Plesset, Note on an approximation treatment for many-electron systems, *Phys. Rev.* **46**, 618.

1939

4 D. R. Hartree, W. Hartree, and B. Swirles, Self-consistent field, including exchange and superposition of configurations, with some results for oxygen, *Phil. Trans. Roy. Soc. (London)* **A238**, 229.

1940–1950

5 M. Kotani, A. Amemiya, E. Ishiguro, and T. Kimura, *Table of molecular integrals,* Maruzen, Tokyo, 1955.

1950

6 S. F. Boys, Electronic wave functions. I. A general method of calculation for the stationary states of any molecular system, *Proc. Roy. Soc. (London)* **A200**, 542.

7 S. F. Boys, Electronic wave functions. II. A calculation for the ground state of the beryllium atom, *Proc. Roy. Soc. (London)* **A201**, 125.

1951

8 K. Ruedenberg, A Study of two-center integrals useful in calculations on molecular structure: II. The two-center exchange integrals, *J. Chem. Phys.* **19**, 1459.

9 C. C. J. Roothaan, New developments in molecular orbital theory, *Rev. Mod. Phys.* **23**, 69

1952

10 G. R. Taylor and R. G. Parr, Superposition of configurations: the helium atom, *Proc. U.S. Natl. Acad. Sci.* **38**, 154.

11 R. G. Parr and B. L. Crawford, National Academy of Sciences conference on quantum mechanical methods in valence theory, *Proc. U.S. Natl. Acad. Sci.* **38**, 547 (Shelter Island).

1954

12 J. A. Pople and R. K. Nesbet, Self-consistent orbitals for radicals, *J. Chem. Phys.* **22**, 571.

1955

13 H. Shull and P. O. Löwdin, Natural spin orbitals for helium, *J. Chem. Phys.* **23**, 1565.

14 P. O. Löwdin, Quantum theory of many-particle systems. I. Physical interpretations by means of density matrices, natural spin-orbitals, and convergence problems in the method of configuration interaction, *Phys. Rev.* **97**, 1474.

15 R. K. Nesbet, Configuration interaction in orbital theories, *Proc. Roy. Soc. (London)* **A230**, 312.

1956

16 S. F. Boys, G. B. Cook, C. M. Reeves, and I. Shavitt, Automatic fundamental calculations of molecular structure, *Nature (Lond.)* **178**, 1207.

1957

17 J. Miller, R. H. Friedman, R. P. Hurst, and F. A. Matsen, Electronic energy of LiH and BeH^+, *J. Chem. Phys.* **27**, 1385.

1958

18 C. L. Pekeris, Ground state of two-electron atoms, *Phys. Rev.* **112**, 1649.

1959

19 P. O. Löwdin, Correlation problem in many-electron quantum mechanics. I. Review of different approaches and discussion of some current ideas, *Adv. Chem. Phys.* **2**, 207.

20 R. S. Mulliken and C. C. J. Roothaan, Broken bottlenecks and the future of molecular quantum mechanics, *Proc. U.S. Natl. Acad. Sci.* **45**, 394.

1960

21 A. D. McLean LCAO-MO-SCF ground state calculations on C_2H_2 and CO_2, *J. Chem. Phys.* **32**, 1595.

22 E. R. Davidson, First excited ${}^1\Sigma_g^+$ state of H_2. A double-minimum problem, *J. Chem. Phys.* **33**, 1577.

23 R. E. Watson, Approximate wavefunctions for atomic Be, *Phys. Rev.* **119**, 170.

24 C. C. J. Roothaan, Self-consistent field theory for open shells of electronic systems, *Rev. Mod. Phys.* **32**, 179.

25 B. J. Ransil, Studies in molecular structure. I. Scope and summary of the diatomic molecule program, *Rev. Mod. Phys.* **32**, 239.

26 R. K. Nesbet, Diatomic molecule project at RIAS and Boston University, *Rev. Mod. Phys.* **32**, 272.

27 S. F. Boys and G. B. Cook, Mathematical problems in the complete quantum predictions of chemical phenomena, *Rev. Mod. Phys.* **32**, 285.

28 J. M. Foster and S. F. Boys, Quantum variational calculations for a range of CH_2 configurations, *Rev. Mod. Phys.* **32**, 305.

1961

29 O. Sinanoglu, Many-electron theory of atoms and molecules, *Proc. U.S. Natl. Acad. Sci.* **47**, 1217.

1962

30 I. Shavitt and M. Karplus, Multicenter integrals in molecular quantum mechanics, *J. Chem. Phys.* **36**, 550.

1963

31 E. Clementi, Correlation energy for atomic systems, *J. Chem. Phys.* **38**, 2248.

32 R. M. Pitzer and W. N. Lipscomb, Calculation of the barrier to internal rotation in ethane, *J. Chem. Phys.* **39**, 1995.

33 C. C. J. Roothaan and P. S. Bagus, Atomic self-consistent field calculations by the expansion method, *Methods in Comp. Phys.* **2**, 47.

34 H. P. Kelly, Correlation effects in atoms, *Phys. Rev.* **131**, 684.

35 C. Edmiston and K. Ruedenberg, Localized atomic and molecular orbitals, *Rev. Mod. Phys.* **35**, 457.

36 R. K. Nesbet, Computer programs for electronic wave-function calculations, *Rev. Mod. Phys.* **35**, 552.

37 M. P. Barnett, Mechanized molecular calculations—the POLYATOM system, *Rev. Mod. Phys.* **35**, 571.

38 S. Hagstrom and H. Shull, The nature of the two-electron chemical bond. III. Natural orbitals for H_2, *Rev. Mod. Phys.* **35**, 624.

1964

39 A. C. Wahl, Analytic self-consistent field wavefunctions and computed properties for homonuclear diatomic molecules, *J. Chem. Phys.* **41**, 2600.

40 M. Krauss, Calculation of the geometrical structure of some AH_n molecules, *J. Res. Nat. Bur. Stand., Sec. A* **68**, 635.

1965

41 E. Clementi, Tables of atomic functions, Supplement to *IBM J. Res. Develop.* **9**, 2.

42 S. Huzinaga, Gaussian-type functions for polyatomic systems. I, *J. Chem. Phys.* **42**, 1293.

43 R. K. Nesbet, Algorithm for diagonalization of large matrices, *J. Chem. Phys.* **43**, 311.

44 J. M. Schulman and J. W. Moskowitz, Preliminary results of a self-consistent-field study of the benzene molecule, *J. Chem. Phys.* **43**, 3287.

45 P. S. Bagus, Self-consistent-field wave functions for hole states of some Ne-like and Ar-like ions, *Phys. Rev.* **139**, A619.

1966

46 G. Das and A. C. Wahl, Extended Hartree–Fock wavefunctions:

optimized valence configurations for H_2 and Li_2, optimized double configurations for F_2, *J. Chem. Phys.* **44**, 87.

47 P. F. Fougere and R. K. Nesbet, Electronic structure of C_2, *J. Chem. Phys.* **44**, 285.

48 J. L. Whitten, Gaussian lobe function expansions of Hartree–Fock solutions for the first-row atoms and ethylene, *J. Chem. Phys.* **44**, 359.

49 H. J. Silverstone and O. Sinanoglu, Many-electron theory of nonclosed-shell atoms and molecules. I. Orbital wavefunction and perturbation theory, *J. Chem. Phys.* **44**, 1899.

50 P. E. Cade, K. D. Sales, and A. C. Wahl, Electronic structure of diatomic molecules. III. A. Hartree–Fock wavefunctions and energy quantities for N_2 ($X\ ^1\Sigma_g^+$) and N_2^+ ($X\ ^2\Sigma_g^+$, $A\ ^2\Pi_u$, $B\ ^2\Sigma_u^+$) molecular ions, *J. Chem. Phys.* **44**, 1973.

51 S. D. Peyerimhoff, R. J. Buenker, and L. C. Allen, Geometry of molecules. I. Wavefunctions for some six- and eight-electron polyhydrides, *J. Chem. Phys.* **45**, 734.

52 C. Edmiston and M. Krauss, Pseudonatural orbitals as a basis for the superposition of configurations. I. He_2^+, *J. Chem. Phys.* **45**, 1833.

53 J. Cizek, On the correlation problem in atomic and molecular systems. Calculation of wavefunction components in Ursell-type expansion using quantum-field theoretical methods, *J. Chem. Phys.* **45**, 4256.

54 C. F. Bender and E. R. Davidson, A natural orbital based energy calculation for helium hydride and lithium hydride, *J. Phys. Chem.* **70**, 2675.

1967

55 A. D. McLean and M. Yoshimine, Tables of linear molecule wave functions, Supplement to *IBM J. Res. Develop.* **12**, 206.

56 F. Grimaldi, A. Lecourt, and C. Moser, The calculation of the electric dipole moment of CO, *Int. J. Quantum Chem. Symp.* **1**, 153.

57 M. Yoshimine and A. D. McLean, Ground states of linear molecules: dissociation energies and dipole moments in the Hartree–Fock approximation, *Int. J. Quantum Chem. Symp.* **1**, 313.

58 A. A. Frost, Floating spherical Gaussian orbital model of molecular structure I. Computational procedure. LiH as an example, *J. Chem. Phys.* **47**, 3707.

59 E. Clementi and J. N. Gayles, Study of the electronic structure of molecules. VII. Inner and outer complex in the NH_4Cl formation from NH_3 and HCl, *J. Chem. Phys.* **47**, 3837.

60 R. K. Nesbet, Atomic Bethe-Goldstone equations. I. The Be atom, *Phys. Rev.* **155**, 51.

1968

61 R. Ahlrichs and W. Kutzelnigg, Direct calculation of approximate natural orbitals and natural expansion coefficients of atomic and molecular electronic wavefunctions. II. Decoupling of the pair equations and calculation of the pair correlation energies for the Be and LiH ground states, *J. Chem. Phys.* **48**, 1819.

62 K. Morokuma and L. Pedersen, Molecular-orbital studies of hydrogen bonds. An *ab initio* calculation for dimeric H_2O, *J. Chem. Phys.* **48**, 3275.

63 W. Kolos and L. Wolniewicz, Improved theoretical ground-state energy of the hydrogen molecule, *J. Chem. Phys.* **49**, 404.

64 E. R. Davidson and C. F. Bender, Correlation energy calculations and unitary transformations for LiH, *J. Chem. Phys.* **49**, 465.

65 J. Gerratt and I. M. Mills, Force constants and dipole-moment derivatives of molecules from perturbed Hartree-Fock calculations. I, *J. Chem. Phys.* **49**, 1719.

66 D. Neumann and J. W. Moskowitz, One-electron properties of near-Hartree-Fock wave functions; I. Water, *J. Chem. Phys.* **49**, 2056.

67 C. F. Bunge, Electronic wave functions for atoms. I. Ground state of Be, *Phys. Rev.* **168**, 92.

68 H. F. Schaefer and F. E. Harris, Metastability of the ^1D state of the nitrogen negative ion, *Phys. Rev. Lett.* **21**, 1561. (First-order wavefunction.)

1969

69 T. H. Dunning, W. J. Hunt, and W. A. Goddard, The theoretical description of the $(\pi\pi^*)$ excited states of ethylene, *Chem. Phys. Lett.* **4**, 147.

70 R. K. Nesbet, T. L. Barr, and E. R. Davidson, Correlation energy of the neon atom, *Chem. Phys. Lett.* **4**, 203.

71 L. C. Snyder and H. Basch, Heats of reaction from self-consistent field energies of closed-shell molecules, *J. Amer. Chem. Soc.* **91**, 2189.

72 R. C. Ladner and W. A. Goddard, Improved quantum theory of many-electron systems. V. The spin-coupling optimized GI method, *J. Chem. Phys.* **51**, 1073.

73 P. Pulay, *Ab initio* calculation of force constants and equilibrium geometries in polyatomic molecules. I. Theory, *Mol. Phys.* **17**, 197.

74 U. Kaldor and F. E. Harris, Spin-optimized self-consistent field wave functions, *Phys. Rev.* **183**, 1.

75 C. F. Bender and E. R. Davidson, Studies in configuration interaction: the first-row diatomic hydrides, *Phys. Rev.* **183**, 21.

1970

76 R. P. Hosteny, R. R. Gilman, T. H. Dunning, A. Pipano, and I. Shavitt, Comparison of Slater and contracted Gaussian basis sets in SCF and CI calculations on H_2O, *Chem. Phys. Lett.* **7**, 325.

77 A. K. Q. Siu and E. R. Davidson, A study of the ground state wave function of carbon monoxide, *Int. J. Quantum Chem.* **4**, 223.

78 C. F. Bender and H. F. Schaefer, New theoretical evidence for the nonlinearity of the triplet ground state of methylene, *J. Amer. Chem. Soc.* **92**, 4984.

79 D. M. Silver, K. Ruedenberg, and E. L. Mehler, Electron correlation and separated pair approximation in diatomic molecules. III. Imidogen, *J. Chem. Phys.* **52**, 1206.

80 M. D. Newton, W. A. Lathan, W. J. Hehre, and J. A. Pople, Self-consistent molecular orbital methods. V. *Ab initio* calculation of equilibrium geometries and quadratic force constants, *J. Chem. Phys.* **52**, 4064.

81 P. J. Bertoncini, G. Das, and A. C. Wahl, Theoretical study of the $^1\Sigma^+$, $^3\Sigma^+$, $^3\Pi$, $^1\Pi$ states of NaLi and the $^2\Sigma^+$ state of $NaLi^+$, *J. Chem. Phys.* **52**, 5112.

82 H. F. Schaefer, New approach to electronic structure calculations for diatomic molecules: application to F_2 and Cl_2, *J. Chem. Phys.* **52**, 6241.

83 A. Bunge, Electronic wave functions for atoms, III. Partition of degenerate spaces and ground state of C, *J. Chem. Phys.* **53**, 20.

84 J. M. Schulman and D. N. Kaufman, Application of many-body perturbation theory to the hydrogen molecule, *J. Chem. Phys.* **53**, 477.

85 T. H. Dunning, Gaussian basis functions for use in molecular calculations. I. Contraction of (9s 5p) atomic basis sets for the first-row atoms, *J. Chem. Phys.* **53**, 2823.

86 H. F. Schaefer, D. R. McLaughlin, F. E. Harris, and B. J. Alder, Calculation of the attractive He pair potential, *Phys. Rev. Lett.* **25**, 988.

87 P. J. Bertoncini and A. C. Wahl, *Ab initio* calculation of the helium–helium $^1\Sigma_g^+$ potential at intermediate and large separations, *Phys. Rev. Lett.* **25**, 991.

1971

88 T. H. Dunning and N. W. Winter, Hartree–Fock calculation of the barrier to internal rotation in hydrogen peroxide, *Chem. Phys. Lett.* **11**, 194.

89 F. Grein and T. C. Chang, Multiconfiguration wavefunctions obtained by application of the generalized Brillouin theorem, *Chem. Phys. Lett.* **12**, 44.

90 W. Meyer, Ionization energies of water from PNO–CI calculations, *Int. J. Quantum Chem. Symp.* **5**, 341.

91 E. Clementi, J. Mehl, and W. von Niessen, Study of the electronic structure of molecules. XII. Hydrogen bridges in the guanine-cytosine pair and in the dimeric form of formic acid, *J. Chem. Phys.* **54**, 508.

92 H. F. Schaefer, *Ab initio* potential curve for the $X^3\Sigma_g^-$ state of O_2, *J. Chem. Phys.* **54**, 2207.

93 S. Rothenberg and H. F. Schaefer, Methane as a numerical experiment for polarization basis function selection, *J. Chem. Phys.* **54**, 2764.

94 J. B. Rose and V. McKoy, Applicability of SCF theory to some open-shell states of CO, N_2 and O_2, *J. Chem. Phys.* **55**, 5435.

1972

95 B. Roos, A new method for large-scale CI calculations, *Chem. Phys. Lett.* **15**, 153.

96 J. A. Horsley, Y. Jean, C. Moser, L. Salem, R. M. Stevens, and J. S. Wright, An organic transition state, *J. Amer. Chem. Soc.* **94**, 279.

97 L. R. Kahn and W. A. Goddard, *Ab initio* effective potentials for use in molecular calculations, *J. Chem. Phys.* **56**, 2685.

98 W. J. Hunt, P. J. Hay, and W. A. Goddard, Self-consistent procedures for generalized valence bond wave functions. Applications H_3, BH, H_2O, C_2H_6, and O_2. *J. Chem. Phys.* **57**, 738.

99 E. Clementi and H. Popkie, Study of the structure of molecular complexes. I. Energy surface of a water molecule in the field of a lithium positive ion, *J. Chem. Phys.* **57**, 1077.

100 C. F. Bender, S. V. O'Neil, P. K. Pearson and H. F. Schaefer, Potential energy surface including electron correlation for $F + H_2 \rightarrow FH + H$: refined linear surface, *Science* **176**, 1412.

1973

101 K. Ruedenberg, R. C. Raffenetti, and R. D. Bardo, Even tempered orbital bases for atoms and molecules, pages 164–9 of *Energy, structure, and reactivity*, editors D. W. Smith and W. B. McRae (Wiley, New York, 1973).

102 J. Rose, T. Shibuya, and V. McKoy, Application of the equations-of-motion method to the excited states of N_2, CO, and C_2H_4, *J. Chem. Phys.* **58**, 74.

103 W. Meyer, PNO–CI studies of electron correlation effects. I. Configuration expansion by means of nonorthogonal orbitals and application to the ground state and ionized states of methane, *J. Chem. Phys.* **58**, 1017.

104 R. C. Raffenetti, General contraction of Gaussian atomic orbitals: core, valence, polarization, and diffuse basis sets; molecular integral evaluation, *J. Chem. Phys.* **58**, 4452.

105 G. C. Lie, J. Hinze, and B. Liu, Valence excited states of CH. I. Potential curves, *J. Chem. Phys.* **59**, 1872.

106 R. M. Pitzer, Electron repulsion integrals and symmetry adapted charge distributions, *J. Chem. Phys.* **59**, 3308.

107 I. Shavitt, C. F. Bender, A. Pipano, and R. P. Hosteny, The iterative calculation of several of the lowest or highest eigenvalues and corresponding eigenvectors of very large symmetric matrices, *J. Comput. Phys.* **11**, 90.

108 M. Yoshimine, Construction of the hamiltonian matrix in large configuration interaction calculations, *J. Comput. Phys.* **11**, 449.

109 U. Kaldor, Many-body perturbation-theory calculations with finite, bound basis sets, *Phys. Rev. A* **7**, 427.

110 M. -M. Coutiere, J. Demuynck, and A. Veillard, Ionization potentials of ferrocene and Koopmans' theorem. An *ab initio* LCAO–MO–SCF calculation, *Theoret. Chim. Acta* **27**, 281.

1974

111 S. R. Langhoff and E. R. Davidson, Configuration interaction calculations on the nitrogen molecule, *Int. J. Quantum Chem.* **8**, 61.

112 P. J. Hay and I. Shavitt, *Ab initio* configuration interaction studies of the π-electron states of benzene, *J. Chem. Phys.* **60**, 2865.

113 J. Paldus, Group theoretical approach to the configuration interaction and perturbation theory calculations for atomic and molecular systems, *J. Chem. Phys.* **61**, 5321.

114 F. Sasaki and M. Yoshimine, Configuration-interaction study of atoms. I. Correlation energies of B, C, N, O, F, and Ne, *Phys. Rev. A* **9**, 17.

1975

115 R. Ahlrichs, H. Lischka, V. Staemmler, and W. Kutzelnigg, PNO-CI (pair natural orbital configuration interaction) and CEPA-PNO (coupled electron pair approximation with pair natural orbitals) calculations of molecular systems. I. Outline of the method for closed-shell states, *J. Chem. Phys.* **62**, 1225.

116 E. A. McCullough, The partial-wave self-consistent-field method for diatomic molecules: computational formalism and results for small molecules, *J. Chem. Phys.* **62**, 3991.

117 C. W. Bauschlicher, D. H. Liskow, C. F. Bender, and H. F. Schaefer, Model studies of chemisorption. Interaction between atomic hydrogen and beryllium clusters, *J. Chem. Phys.* **62**, 4815.

118 E. R. Davidson, The iterative calculation of a few of the lowest eigenvalues and corresponding eigenvectors of large real-symmetric matrices, *J. Comput. Phys.* **17**, 87.

119 G. H. F. Diercksen, W. P. Kraemer, and B. O. Roos, SCF–CI studies of correlation effects on hydrogen bonding and ion hydration. The systems: H_2O, $H^+ \cdot H_2O$, $Li^+ \cdot H_2O$, $F^- \cdot H_2O$, and $H_2O \cdot H_2O$, *Theoret. Chim. Acta.* **36**, 249.

120 R. Ahlrichs and F. Driessler, Direct determination of pair natural orbitals. A new method to solve the multi-configuration Hartree-Fock problem for two-electron wave functions, *Theoret. Chim. Acta* **36**, 275.

1976

121 H. F. Schaefer and W. H. Miller, Large scale scientific computation via minicomputer, *Computers and Chemistry* **1**, 85.

122 S. R. Langhoff and E. R. Davidson, *Ab initio* evaluation of the fine structure and radiative lifetime of the $^3A_2$ ($n \rightarrow \pi^*$) state of formaldehyde, *J. Chem. Phys.* **64**, 4699.

123 M. Dupuis, J. Rys, and H. F. King, Evaluation of molecular integrals over Gaussian basis functions, *J. Chem. Phys.* **65**, 111.

124 P. Rosmus and W. Meyer, Spectroscopic constants and the dipole moment functions for the $^1\Sigma^+$ ground state of NaLi, *J. Chem. Phys.* **65**, 492.

125 C. E. Dykstra, H. F. Schaefer, and W. Meyer, A theory of self-consistent electron pairs. Computational methods and preliminary applications, *J. Chem. Phys.* **65**, 2740.

126 L. G. Yaffe and W. A. Goddard, Orbital optimization in electronic wave functions; equations for quadratic and cubic convergence of general multiconfiguration wave functions, *Phys. Rev. A* **13**, 1682.

127 C. F. Bunge, Accurate determination of the total electronic energy of the Be ground state, *Phys. Rev. A* **14**, 1965.

1977

128 I. Shavitt, Graph theoretical concepts for the unitary group approach to the many-electron correlation problem, *Int. J. Quantum Chem. Symp.* **11**, 131.

129 W. von Niessen, G. H. F. Diercksen, and L. S. Cederbaum, On the accuracy of ionization potentials calculated by Green's functions, *J. Chem. Phys.* **67**, 4124.

130 R. P. Saxon and B. Liu, *Ab initio* configuration interaction study of the valence states of O_2, *J. Chem. Phys.* **67**, 5432.

1978

131 J. E. Gready, G. B. Bacskay, and N. S. Hush, Finite-field method calculations. IV. Higher-order moments, dipole moment gradients, polarisability gradients, and field-induced shifts in molecular properties: application to N_2, CO, CN^-, HCN, and HNC, *Chem. Phys.* **31**, 467.

132 J. A. Pople, R. Krishnan, H. B. Schlegel, and J. S. Binkley, Electron correlation theories and their application to the study of simple reaction potential surfaces, *Int. J. Quantum Chem.* **14**, 545.

133 R. J. Bartlett and G. D. Purvis, Many-body perturbation theory, coupled-pair many-electron theory, and the importance of quadruple excitations for the correlation problem, *Int. J. Quantum Chem.* **14**, 561.

134 P. Siegbahn and B. Liu, An accurate three-dimensional potential energy surface for H_3, *J. Chem. Phys.* **68**, 2457.

135 R. J. Buenker, S. D. Peyerimhoff, and W. Butscher, Applicability of the multi-reference double-excitation CI (MRD–CI) method to the calculation of electronic wavefunctions and comparison with related techniques, *Mol. Phys.* **35**, 771.

1979

136 J. A. Pople, R. Krishnan, H. B. Schlegel, and J. S. Binkley, Derivative studies in Hartree-Fock and Møller–Plesset theories, *Int. J. Quantum Chem. Symp.* **13**, 225.

137 B. R. Brooks and H. F. Schaefer, The graphical unitary group approach to the electron correlation problem. Methods and preliminary applications, *J. Chem. Phys.* **70**, 5092.

138 P. E. M. Siegbahn, Generalizations of the direct CI method based on the graphical unitary group approach. I. Single replacements from a complete CI root function of any spin, first order wave functions, *J. Chem. Phys.* **70**, 5391.

1980

139 B. O. Roos, P. R. Taylor, and P. E. M. Siegbahn, A complete active space SCF method (CASSCF) using a density matrix formulated super-CI approach, *Chem. Phys.* **48**, 157.

140 B. Liu and A. D. McLean, *Ab initio* potential curve for $Be_2\,(^1\Sigma_g^+)$ from the interacting correlated fragments method, *J. Chem. Phys.* 72, 3418.

141 R. Krishnan, M. J. Frisch, and J. A. Pople, Contribution of triple substitutions to the electron correlation energy in fourth order perturbation theory, *J. Chem. Phys.* **72**, 4244.

142 B. R. Brooks, W. D. Laidig, P. Saxe, J. D. Goddard, Y. Yamaguchi, and H. F. Schaefer, Analytic gradients from correlated wave functions via the two-particle density matrix and the unitary group approach, *J. Chem. Phys.* **72**, 4652.

143 R. Krishnan, H. B. Schlegel, and J. A. Pople, Derivative studies in configuration-interaction theory, *J. Chem. Phys.* 72, 4654.

144 B. H. Lengsfield, General second order MCSCF theory : a density matrix directed algorithm, *J. Chem. Phys.* **73**, 382.

1982

145 Y. Osamura, Y. Yamaguchi, P. Saxe, M. A. Vincent, J. F. Gaw, and H. F. Schaefer, Unified theoretical treatment of analytic first and second energy derivatives in open-shell Hartree–Fock theory, *Chem. Phys.* **72**, 131.

146 G. D. Purvis and R. J. Bartlett, A full coupled-cluster singles and doubles model: the inclusion of disconnected triples, *J. Chem. Phys.* **76**, 1910.

147 Y. Osamura, Y. Yamaguchi, and H. F. Schaefer, Generalization of analytic configuration interaction (CI) gradient techniques for potential energy hypersurfaces, including a solution to the coupled perturbed Hartree–Fock equations for multiconfiguration SCF molecular wave functions, *J. Chem. Phys.* **77**, 383.

148 J. Almlöf, K. Faegri, and K. Korsell, Principles for a direct SCF approach to LCAO–MO *ab initio* calculations, *J. Comput. Chem.* **3**, 385.

1983

149 V. R. Saunders and J. H. van Lenthe, The direct CI method. A detailed analysis, *Mol. Phys.* **48**, 923.

1

E. A. HYLLERAAS

Über den Grundzustand des Heliumatoms

Z. Physik. **48**, 469 (1928)

This is the first of a series of three classic papers published by Hylleraas in 1928, 1929, and 1930, respectively. In this the first paper Hylleraas introduced the method of superposition of configurations, or configuration interaction (CI), as it is more commonly known today. The CI method, in which the wavefunction is variationally determined as a linear combination of (antisymmetrized) spin-orbital products, remains in 1984 the most widely adopted approach to the treatment of correlated wavefunctions, i.e., those going beyond the Hartree-Fock approximation. In this sense Hylleraas may be considered the father of much of modern electronic structure theory. In Hylleraas's second paper,[a] he introduced the interelectronic distance r_{12} explicitly in the wavefunction. Although the latter method gave remarkably close agreement with experiment, as demonstrated for the He isoelectronic series in the 1930 paper,[b] it is even now (more than 50 years later) not readily applicable to atoms with more than a few electrons.

[a] E. A. Hylleraas, Neue Berechnung der Energie des Heliums in Grundzustande, sowie des tiefsten Terms von Orthohelium, *Z. Physik* **54**, 347 (1929).
[b] E. A. Hylleraas, Über den Grundterm der Zweielektronenprobleme von H^-, He, Li^+, Be^{++}usw., *Z. Physik* **65**, 209 (1930).

2

H. M. JAMES & A. S. COOLIDGE

The ground state of the hydrogen molecule

J. Chem. Phys. **1**, 825 (1933)

This classic paper did for the hydrogen molecule what Hylleraas had done for the helium atom three years earlier. James and Coolidge followed Hylleraas in his explicit incorporation of the distance between the two electrons in the wavefunction. For H_2, of course, the resulting two-electron integrals are much more difficult to evaluate than are those for the helium atom. The total energy obtained by James and Coolidge was −1.173 47 hartrees, compared to the exact non-relativistic result, −1.174 47 hartrees, and the dissociation energy was D_e = 4.70 eV, in close agreement with experiment, 4.75 eV. This paper was a smashing success in proving that Schrödinger's equation was trustworthy for molecules as well as atoms. However, the paper may have unintentionally served a regressive purpose, in that it provided discouragement for the much more widely applicable Hartree–Fock and configuration interaction (CI) methods. Many years passed[a,b] before the more general CI approach was to provide variational wavefunctions comparable in accuracy to that of James and Coolidge.

[a] S. Hagstrom and H. Shull, The nature of the two-electron chemical bond. III. Natural orbitals for H_2, *Rev. Mod. Phys.* **35**, 624 (1963).

[b] The first conventional CI wave function to yield an energy (−1.173 70 hartrees) lower than that of James and Coolidge was reported by B. Liu, *Ab initio* potential energy curve for linear H_3, *J. Chem. Phys.* **58**, 1925 (1973). Liu carried out full CI with a Slater basis set of size 4s 3p 2d on each H atom.

3

C. MØLLER & M. S. PLESSET

Note on an approximation treatment for many-electron systems

Phys. Rev. **46**, 618 (1934)

This 'note' is in fact five pages long and is one of the key papers in the development of molecular quantum mechanics. Here Møller and Plesset develop a perturbation theory in which the Hartree–Fock wavefunction is taken as the zero-order approximation to the exact theory. That is, the difference between the exact hamiltonian and Hartree–Fock hamiltonian is regarded as a small perturbation. Møller and Plesset found that the first-order correction to the energy vanishes and derived a general expression for the second-order energy. The development of efficient methods, both by diagrammatic many-body perturbation theory[a] and the direct algebraic evaluation of the third- and fourth-order energies, has led to a resurgence of interest in techniques based on the ideas of this classic paper. The well-known theorem of Møller and Plesset (which may also be viewed as a corollary of Brillouin's theorem[b]) states that the Hartree–Fock electronic density (and indirectly other one-electron properties such as the dipole moment) are correct through the first order of perturbation theory.

[a] K. A. Breuckner, Many-body problem for strongly interacting particles. II. Linked cluster expansion, *Phys. Rev.* **100**, 36 (1955).
[b] L. Brillouin, *La méthode du champ self-consistent*, Actualités Sci. Ind. No. 71 (Herman & Cie, Paris, 1933).

4

D. R. HARTREE, W. HARTREE, & B. SWIRLES

Self-consistent field, including exchange and superposition of configurations, with some results for oxygen

Phil. Trans. Roy. Soc. (London) **A238**, 229 (1939)

It is widely known that D. R. Hartree, as well as being the discoverer of the Hartree equations (which neglect exchange) was also the pioneer (with his co-workers, especially his father, W. Hartree) in the numerical determination of Hartree and Hartree–Fock wavefunctions.[a] In this paper Hartree and his father and Bertha Swirles reported the first multiconfiguration self-consistent-field (MCSCF) wavefunctions for any system. It was of course also necessary for Hartree to work out the formalism required for the specific types of MCSCF wavefunction needed. Unknown to Hartree, the first reference to the notion of MCSCF was apparently given five years earlier in a book by Frenkel.[b] The actual system considered by Hartree was the ^{2}P state of O^+ (this is the second excited state of O^+; the ground and first excited states are of ^{4}S and ^{2}D symmetry, respectively) and the resulting normalized two-configuration SCF wave function was

$$\psi = \begin{array}{l} 0.98\ 1s^2 2s^2 2p^3 \\ +0.20\ 1s^2 2p^5 \end{array}$$

An important qualitative result was that the MCSCF orbitals were very similar to those obtained via the ordinary one-configuration SCF method.

[a] D. R. Hartree, The wave mechanics of an atom with a non-coulomb central field. I. Theory and methods. II. Some results and discussion. III. Term values and intensities in optical spectra, *Proc. Cambridge Phil. Soc.* **24**, 89, 111, 426 (1928).
[b] J. Frenkel, *Wave mechanics, advanced general theory* (Clarendon Press, Oxford, 1934).

5

M. KOTANI, A. AMEMIYA, E. ISHIGURO, & T. KIMURA

Table of molecular integrals

Maruzen, Tokyo, 1955

The 1940s were, in a certain sense, the 'dark ages' of *ab initio* electronic structure theory. It was not apparent how to extend the highly reliable results of Hylleraas for He and James and Coolidge for H_2 to larger atoms and molecules. Although the Hartree-Fock and multiconfiguration Hartree-Fock formalisms were available, the application of these *ab initio* methods even to molecules as small as water seemed impossible at the time. One bright spot during the late 1930s and the 1940s was the group of Kotani and co-workers in Tokyo. They developed practical methods for the calculation of one- and two-electron integrals required for valence bond and molecular orbital treatments of diatomic molecules. Kotani also made extensive use of Serber[a] and Yamanouchi's[b] earlier work on spin eigenfunctions in laying out a systematic approach for the application of valence bond theory. The research of the Kotani team during the 1940s was summarized in the book *Tables of molecular integrals*, which proved to be a very valuable volume for workers in the field.

[a] R. Serber, Extension of the Dirac vector model to include several configurations, *Phys. Rev.* **45**, 461 (1934).
[b] T. Yamanouchi, Calculations of atomic energy levels, *Proc. Phys. -Math. Soc. Japan* **18**, 623 (1936).

6

S. F. BOYS

Electronic wave functions. I. A general method of calculation for the stationary states of any molecular system

Proc. Roy. Soc. (London) **A200**, 542 (1950)

Since the exact solutions of the Schrödinger equation for the hydrogen atom are linear combinations of Slater functions or Slater-type orbitals (STO), of the general form $r^n e^{-\zeta r}$, it had almost universally been assumed that this type of basis function should be used in valence bond and molecular orbital wavefunctions. However, it was early discovered that four-centre, two-electron integrals involving Slater functions are extraordinarily difficult to evaluate. Despite heroic efforts along these lines over the years,[a] today STOs are primarily restricted as basis functions to the study of diatomic (and sometimes linear) molecules. Thus the present paper by Boys, introducing gaussian functions $x^l y^m z^n e^{-\alpha r^2}$, is central to most of the progress which has occurred in electronic structure theory during the past 15 years. Gaussian functions are inherently less effective, on a one-to-one basis, than Slaters in describing molecular orbitals. Thus, Boys's genius lay in recognizing that all resulting gaussian integrals could be evaluated rapidly using analytic techniques, so that gaussians could be competitive on a several-to-one basis. Furthermore Boys presented the requisite formulae for these one- and two-electron integrals over gaussian functions.[b]

[a] For early work, see M. P. Barnett and C. A. Coulson, The evaluation of integrals occurring in the theory of molecular structure, *Trans. Roy. Soc. (London)* **A243**, 221 (1951).

[b] An alternate, more explicit set of gaussian integral formulae is given by H. Taketa, S. Huzinaga, and K. O-Ohata, Gaussian-expansion methods for molecular integrals, *J. Phys. Soc. Japan* **21**, 2313 (1966).

7

S. F. BOYS

Electronic wave functions.
II. A calculation for the ground state of the beryllium atom

Proc. Roy. Soc. (London) **A201**, 125 (1950)

In the Boys paper just considered, the second half is devoted to one of the earliest systematic discussions of the general method of configuration interaction (CI).[a] This paper makes the first application of those ideas. Interestingly, Boys used Slater functions in this calculation on the beryllium atom, for the obvious reason that in the atomic case the resulting integrals were readily evaluated. Using orthogonal linear combinations of STOs as atomic orbitals (1s, 2s, 3s, and 2p), a ten configuration variational wavefunction was determined. This was perhaps the first *ab initio* CI wave function for a system of more than three electrons.[b] Boys obtained an energy of −14.6220 hartrees, significantly below the −14.5730 we now know to be the Hartree-Fock limit. Since less work was required than using the standard purely numerical Hartree-Fock method, Boys was justifiably pleased with his results.

[a] It should be noted that in the context of semi-empirical theory, Craig made important advances of a related nature at about the same time. See, for example, D. P. Craig, Configurational interaction in the molecular orbital theory. A higher approximation in the nonempirical method, *Proc. Roy. Soc. (London)* **A200**, 474 (1950).

[b] Nonempirical valence bond (equivalent to a minimum basis full CI) studies of H_3 had been reported rather early by J. O. Hirschfelder, Energy of the triatomic hydrogen molecule and ion: V, *J. Chem. Phys*, **6**, 795 (1938).

8

K. RUEDENBERG

A study of two-center integrals useful in calculations on molecular structure. II. The two-center exchange integrals

J. Chem. Phys. **19**, 1459 (1951)

This paper by Ruedenberg opened the way for the use of Slater basis functions in quantum mechanical studies of linear molecules.[a] The most intransigent of the electron repulsion integrals was the two-centre exchange integral

$$\int\int \chi_{Ai}(1)^* \, \chi_{Bj}(2)^* \frac{1}{r_{12}} \chi_{Bk}(1) \, \chi_{Al}(2) \, dv_1 \, dv_2,$$

where A, B indicate the nuclei and i, j, k, l indicate atomic orbitals on each nucleus. Ruedenberg exploited the use of charge distributions $\Omega(1) = \chi_a(1)\,\chi_b(1)$, which collect the part of the integral referring to electron 1, and the use of the elliptical coordinates ξ and η, but his most critical insight was a rearrangement of the integrand and subsequent integration by parts to obtain a much more tractable formula. For Slater functions of the types 1s, 2s, 2p, 3s, 3p, and 3d, Ruedenberg then showed that there were only 65 unique types of charge distributions and these give rise to 47 types of exchange integrals, which were the immediate object of early *ab initio* studies.[b]

[a] An important later paper in the spirit of Ruedenberg's work here is that of F. E. Harris, Molecular orbital studies of diatomic molecules. I. Method of computation for single configurations of heteronuclear systems, *J. Chem. Phys.* **32**, 3 (1960).

[b] Experience gained in the 1950s and 1960s has naturally refined the techniques for the evaluation of diatomic Slater integrals. See E. L. Mehler and K. Ruedenberg, Two-center exchange integrals between Slater-type atomic orbitals, *J. Chem. Phys.* **50**, 2575 (1969).

9

C. C. J. ROOTHAAN

New developments in molecular orbital theory

Rev. Mod. Phys. **23**, 69 (1951)

In this classic paper, Roothaan set out a rigorous mathematical formalism for the solution of the self-consistent-field (SCF) equations in terms of a finite basis set of analytic (one-electron) functions.[a] Since Roothaan's paper, this method has variously been referred to as the analytic Hartree-Fock, Hartree-Fock-Roothaan, or matrix Hartree-Fock method. As one's finite basis set is increased in size, of course, the SCF wavefunction obtained in this way for a particular atomic or molecular system will ultimately approach the true Hartree-Fock wavefunction, i.e., the Hartree-Fock limit. Although the method was only formulated for closed shells, there is a lovely discussion of the use of point group symmetry to simplify the matrix Hartree-Fock equations. This paper also set out the philosophical framework that was to characterize the herculean efforts of the University of Chicago group over the next fifteen years. Roothaan's belief was that first priority in electronic structure theory should go towards solving these analytic Hartree-Fock equations; then one could ponder whether or not the treatment of electron correlation was necessary. Most theoretical chemists today accept the validity of this general perspective.

[a] Related work was published independently by G. G. Hall, The molecular orbital theory of chemical valency. VIII. A method of calculating ionization potentials, *Proc. Roy. Soc. (London)* **A205**, 541 (1951).

10

G. R. TAYLOR & R. G. PARR

Superposition of configurations: the helium atom

Proc. U.S. Natl. Acad. Sci. **38**, 154 (1952)

Taylor and Parr's paper marked the realization that the configuration interaction (CI) method was here to stay. They appreciated that even though the CI method in 1952 was not competitive with the Hylleraas techniques for the He atom, that CI was nonetheless the method of the future for larger systems. Little could they know that thirty years later CI techniques would be able to variationally treat not ten but *one million* configurations.[a] Taylor and Parr's best wavefunction included the four configurations $1s1s'$, $2p^2$, $3d^2$, and $4f^2$ and accounted for 88 per cent of the correlation energy. Their work is characterized by a wealth of physical insight, including one of the earliest precise definitions of the correlation energy, the difference between the Hartree-Fock and experimental energies. Taylor and Parr also differentiated between radial (that obtained using purely radial (1s, 2s, 3s, . . .) basis functions) and angular correlation and estimated their contributions to be 38 per cent and 62 per cent, respectively.

a P. Saxe, D. J. Fox, H. F. Schaefer, and N. C. Handy, The shape-driven graphical unitary group approach to the electron correlation problem. Application to the ethylene molecule, *J. Chem. Phys.* **77**, 5584 (1982).

11

R. G. PARR & B. L. CRAWFORD

National Academy of Sciences conference on quantum mechanical methods in valence theory

Proc. U.S. Natl. Acad. Sci. **38**, 547 (1952)

The Shelter Island Conference has been cited as the first American conference on theoretical chemistry[a] and had a considerable influence on the course of molecular quantum mechanics. Among the 25 participants were (in addition to authors Parr and Crawford) M. P. Barnett, C. A. Coulson, H. Eyring, J. O. Hirschfelder, M. Kotani, J. E. Lennard-Jones, P. O. Löwdin, J. E. Mayer, W. Moffitt, R. S. Mulliken, K. S. Pitzer, C. C. J. Roothaan, K. Ruedenberg, H. Shull, J. C. Slater, and J. H. Van Vleck. The need felt for reliable methods for the computation of molecular integrals was expressed in the abstract. Other contributions summarized in the Parr–Crawford paper include mention of Moffitt's method of 'atoms in molecules'. However the futuristic aspect of this conference was its emphasis on the use of automatic computing machinery (IBM is cited twice) to solve the problems of electronic structure theory. In this sense the Shelter Island Conference marked the emergence of theoretical chemistry from the 'dark ages' of the past decade or more.

[a] J. C. Light, American Conference on Theoretical Chemistry. Introductory remarks, *J. Phys. Chem.* **86**, 2111 (1982).

12

J. A. POPLE & R. K. NESBET

Self-consistent orbitals for radicals

J. Chem. Phys. **22**, 571 (1954)

As noted earlier, Roothaan's 1951 paper on the solution of the SCF equations within a finite, analytic basis set was suitable only for closed-shell ground states. There was thus a need for a method of comparable simplicity, but applicable to the majority of open-shell systems. This background set the stage for Pople and Nesbet's introduction of the unrestricted Hartree–Fock (UHF) method. In this method the spatial orbitals associated with α and β spin functions are allowed to be distinct. It follows that for the lithium atom[a] the UHF wavefunction is of the form

$$1s(1)\alpha(1)\ 1s'(2)\beta(2)\ 2s(3)\alpha(3),$$

with the 1s and 1s$'$ functions allowed to be nonorthogonal and spatially distinct. A major problem associated with the UHF method is that it does not provide an exact eigenfunction of the spin operator $\mathbf{S}^2$. Although the UHF method was somewhat neglected between 1960 and 1975 for this reason, it has recently enjoyed a resurgence of interest due to the simplicity of carrying out high orders of perturbation theory with respect to a UHF reference function.[b]

[a] For one of the first *ab initio* applications of the UHF method, see G. W. Pratt, Unrestricted Hartree–Fock method, *Phys. Rev.* **102**, 1303 (1956).

[b] J. A. Pople, J. S. Binkley, and R. Seeger, Theoretical models incorporating electron correlation, *Int. J. Quantum Chem. Symp.* **10**, 1 (1976).

13

H. SHULL & P. O. LÖWDIN

Natural spin orbitals for helium

J. Chem. Phys. **23**, 1565 (1955)

One of the remarkable aspects of Löwdin's classic 1955 paper (to be discussed shortly) was the proof that, in a certain sense 'the introduction of natural spin-orbitals leads . . . to a configurational expansion of most rapid convergence.' The first practical demonstration of the validity of this proof was given less than one year later by Shull and Löwdin in their study of the helium atom. With a simple basis set of three orbitals (1s, 2s, and 3s Laguerre functions), they carried out a full CI, consisting of the six configurations $1s^2$, $2s^2$, $3s^2$, 1s2s, 1s3s, and 2s3s. When the first-order density matrix was diagonalized and the calculations repeated using as atomic orbitals the natural orbitals, the last three configurations were found to have coefficients of zero in the CI. More generally, for two-electron closed-shell systems the use of natural orbitals reduces the number of contributing configurations from $n(n + 1)/2$ to n. Finally, Shull and Löwdin noted that the $2s^2$ configuration is much more important than $3s^2$ and therefore exploited the equivalence[a,b] between the two-configuration wavefunction $1s^2 - \alpha^2 2s^2$ and the single configuration $(1s - \alpha 2s, 1s + \alpha 2s)$, based on non-orthogonal orbitals.

a Shull and Löwdin cite in this regard C. A. Coulson and I. Fischer, Notes on the molecular orbital treatment of the hydrogen molecule, *Phil. Mag.* **40**, 386 (1949).

b This equivalence provides the basis for the generalized valence bond interpretation of multiconfiguration wavefunctions. See F. W. Bobrowicz and W. A. Goddard, The self-consistent field equations for generalized valence bond and open-shell Hartree-Fock wave functions, pages 79-127 of Volume 3, *Modern theoretical chemistry*, editor H. F. Schaefer (Plenum, New York, 1977).

14

P. O. LÖWDIN

Quantum theory of many-particle systems. I. Physical interpretations by means of density matrices, natural spin-orbitals, and convergence problems in the method of configuration interaction

Phys. Rev. **97**, 1474 (1955)

This is one of the very few formal papers since Schrödinger to justify citation in the present rather specialized volume. The stated goal of Löwdin's paper was interpretive in nature, namely to provide a simple physical picture of molecular electronic structure. It seems fair to say that the concept of natural orbitals (NOs) has been remarkably successful in achieving that goal. The structure of an arbitrarily complicated CI wavefunction can be transparently revealed in terms of its natural orbitals and NO occupation numbers. The chemist need only stretch his imagination slightly to appreciate that in more exact theoretical treatments these occupation numbers are no longer integer (as in Hartree–Fock theory) but take on non-integer values. In addition the natural orbitals allow a simple quantum mechanical interpretation of predicted one-electron properties, such as the dipole moment. Finally, and equally importantly, these concepts have inspired much fruitful research on techniques for the actual determination of correlated wavefunctions.[a]

[a] E. R. Davidson, *Reduced density matrices in chemistry* (Academic Press, New York, 1976).

15

R. K. NESBET

Configuration interaction in orbital theories

Proc. Roy. Soc. (London) **A230**, 312 (1955)

This paper presents a major part of Nesbet's thesis research at Cambridge University under the direction of S. F. Boys. Thus it is not surprising that the paper may perhaps best be viewed as a continuation of Boys's first 1950 paper. Perhaps the most important aspect of this paper involves the determination of hamiltonian matrix elements

$$H_{ij} = \int \Phi_i^* H \Phi_j \, d\tau$$

occuring in configuration interaction studies. In the above equation Φ_i and Φ_j are configurations appearing in the wavefunction. Nesbet showed that one need not explicitly consider every spin orbital occupied[a] in Φ_i or Φ_j to evaluate the matrix element H_{ij}. By considering how configurations i and j differ from the (Hartree-Fock) reference configuration Φ_0, a much simpler formalism is obtained. Many of the more recent advances (for example the direct CI method) in the method of configuration interaction have built upon this insight due to Nesbet.

[a] A review of the traditional procedure for obtaining these matrix elements is given in Chapter 12 of J. C. Slater, *Quantum theory of atomic structure*, Volume I (McGraw-Hill, 1960).

16

S. F. BOYS, G. B. COOK, C. M. REEVES, & I. SHAVITT

Automatic fundamental calculations of molecular structure

Nature (Lond.) **178**, 1207 (1956)

In this paper, Boys and his students reported the complete implementation, on the EDSAC electronic computer, of the mathematical formalism set out in his first 1950 paper. The power of Boys's approach was demonstrated by his report of three of the most accurate wave functions then known for molecular systems with more than two electrons. For BH 23 configuration wavefunctions were reported, for H_2O 30 configurations were included,[a] and for the H_3 system the transition state region for the $H + H_2$ exchange reaction was pursued using 35 configurations. In all three cases stationary point geometries and force constants were determined, not just single energies at assumed geometries. The beauty of the method lay not only with the specific theoretical predictions made, but also with the satisfaction that an 'experimental apparatus', had been constructed that would eliminate the incredibly tedious hand computations that had previously plagued this important part of theoretical chemistry.

a A minimum basis set SCF wave function for H_2O (with three-centre integrals approximated) was reported by F. O. Ellison and H. Shull, Molecular calculations. I. LCAO MO self-consistent field treatment of the ground state of H_2O, *J. Chem. Phys.* **23**, 2348 (1955).

17

J. MILLER, R. H. FRIEDMAN, R. P. HURST, & F. A. MATSEN

Electronic energy of LiH and BeH^+

J. Chem. Phys. **27**, 1385 (1957)

Another set of automatic programs, restricted to diatomic molecules but capable of higher absolute accuracy than those of Boys, was established by Matsen's group at the University of Texas on the IBM 650 computer. The present paper, by Miller, Friedman, Hurst, and Matsen, is thus one of the earliest successful post-Hartree–Fock studies of a diatomic molecule. Matsen used a Slater function basis of 1s, 2s, and 2pσ on Li in conjunction with a single 1s function on the hydrogen atom. For LiH, four choices of the orbital exponents were made, and it was concluded that those from Slater's rules[a] gave the best results. With this non-orthogonal basis set, twenty valence bond structures are possible and all were included in the final CI. Later VB CI calculations[b] on lithium hydride in Matsen's group yielded the lowest energy available for that molecule prior to Bender and Davidson's introduction of the iterative natural orbital method.

a J. C. Slater, Atomic shielding constants, *Phys. Rev.* **36**, 57 (1930).
b J. C. Browne and F. A. Matsen, Quantum-mechanical calculations for the electric field gradients and other electronic properties of lithium hydride: the use of mixed basis sets, *Phys. Rev.* **135**, A1227 (1964).

18

C. L. PEKERIS

Ground state of two-electron atoms

Phys. Rev. **112**, 1649 (1958)

In this paper Pekeris reports what is for practical purposes an exact solution of Schrödinger's Equation for the non-relativistic helium atom. Following Hylleraas, Pekeris introduced the interelectronic distance r_{12} directly into the wavefunction. Using a 214-term expansion in conjunction with a special program written for the WEIZAC electronic computer, an absolute accuracy of 0.01 cm^{-1} (5×10^{-8} hartree) was obtained. Corrections for nuclear motion, relativity, and the Lamb shift led eventually to a value of 198 310.67 cm^{-1} for the ionization potential of He, in agreement with Herzberg's experimental value 198 310.82 ± 0.15 cm^{-1}. More recent calculations by Pekeris[a] have now achieved an absolute accuracy of 1×10^{-12} hartree in the non-relativistic energy. Hence our conclusion that the non-relativistic He atom may be considered a solved problem.

[a] K. Frankowski and C. L. Pekeris, Logarithmic terms in the wave functions of the ground state of two-electron atoms, *Phys. Rev.* **146**, 46 (1966).

19

P. O. LÖWDIN

Correlation problem in many-electron quantum mechanics. I. Review of different approaches and discussion of some current ideas

Adv. Chem. Phys. **2**, 207 (1959)

This review article by Löwdin was almost immediately recognized as a landmark and in this author's opinion should continue to be required reading for beginning graduate students in molecular quantum mechanics. Particularly pertinent to the present commentary is the discussion of Section III, 'methods for treating electron correlation', tracing developments from Hylleraas's three classic papers to 1958. Not widely appreciated at that time was the generality of the notion of 'different orbitals for different spins', although some aspects of this method are implicit in the earlier cited papers of Taylor and Parr and of Shull and Löwdin and the book by Kotani. Löwdin noted that the energy of a UHF wavefunction (which is not an eigenfunction of $\mathbf{S}^2$) may in general be improved upon (i.e. lowered) by application of a spin projection operator, yielding of course a multideterminant wavefunction. Much to be preferred, however, is the 'extended Hartree-Fock' method,[a] wherein the orbital optimization is carried out after application of the spin projection operator.

[a] P. O. Löwdin, Quantum theory of many-particle systems. III. Extension of the Hartree-Fock scheme to include degenerate systems and correlation effects, *Phys. Rev.* **97**, 1509 (1955).

20

R. S. MULLIKEN & C. C. J. ROOTHAAN

Broken bottlenecks and the future of molecular quantum mechanics

Proc. U.S. Natl. Acad. Sci. **45**, 394 (1959)

This is one of the few important papers in electronic structure theory of a decidedly philosophical nature.[a] The immediate reason for the paper was the completion, at Chicago, of completely automatic computer programs for SCF studies of diatomic molecules.[b] Thus it became possible to repeat in 35 minutes of computer time a study of the N_2 molecule that had recently required[c] three man-years of painstaking effort. However, the Mulliken-Roothaan paper also provides a retrospective view of the first thirty years of quantum chemistry and looks far into the future. Specifically they state that 'Looking toward the future, it seems certain that colossal rewards lie ahead from large-scale quantum-mechanical calculations of the structure of matter.' From our perspective, it would appear that in 1982 this high calling has yet to be attained, although great progress toward that goal can be realistically claimed. We are confident that by the year 2000, essentially all fields of chemistry will acknowledge the accuracy of Mulliken and Roothaan's prophecy.

[a] For an important later paper in the same spirit, see R. G. Parr, The description of molecular structure, *Proc. U.S. Natl. Acad. Sci.* **72**, 763 (1975).
[b] See also C. C. J. Roothaan, Evaluation of molecular integrals by digital computer, *J. Chem. Phys.* **28**, 982 (1958).
[c] C. W. Scherr, SCF LCAO MO study of N_2, *J. Chem. Phys.* **23**, 569 (1955).

21

A. D. McLEAN

LCAO–MO–SCF ground state calculations on C_2H_2 and CO_2

J. Chem. Phys. **32**, 1595 (1960)

The first completely rigorous (no integral approximations whatever) studies of polyatomic molecules using Slater functions were reported here by McLean. His acetylene and carbon dioxide wavefunctions resulted from a system of automatic programs put together by the University of Chicago group for the UNIVAC 1103A digital computer. A minimum basis set was used, with orbital exponents ζ optimized for the isolated atoms C, O, and H. Great care was exercised to guarantee a precision of 0.0001 hartree (0.06 kcal) in the total energies. Two hours of computer time were required for the acetylene wavefunctions (12 basis functions), and three hours for CO_2 (15 basis functions). For the ionization potentials, theory was found to be 0.61 eV too high for acetylene and 1.81 eV too low for CO_2. Even before this paper was published, McLean was able to indicate that further work on C_2H_2 and CO_2 with larger basis sets was in progress.[a]

[a] A. D. McLean, Extended basis-set LCSTO–MO–SCF calculations on the ground state of carbon dioxide, *J. Chem. Phys.* **38**, 1347 (1963).

22

E. R. DAVIDSON

First excited ${}^1\Sigma_g^+$ state of H_2. A double-minimum problem

J. Chem. Phys. **33**, 1577 (1960)

This is part of Davidson's thesis, completed in the research group of Harrison Shull at Indiana University. It was perhaps the earliest *ab initio* explanation of an otherwise confusing set of spectroscopic observations. Although the first excited ${}^1\Sigma_g^+$ state of H_2 was thought in 1960 to be fairly well known experimentally, the vibration-rotation levels associated with this state showed many irregularities in spacing which were not easily explained. In approaching this problem, Davidson predicted the shape of the potential curve using a 20 configuration variational wavefunction constructed from elliptical basis functions. Davidson also was one of the first to apply the Hylleraas-Undheim-MacDonald theorem,[a,b] to molecular systems in the framework of modern *ab initio* methods. In the present context the HUM theorem guarantees that the second root of the ${}^1\Sigma_g^+$ secular equation provides an upper bound to the true 2 ${}^1\Sigma_g^+$ energy. Davidson's discovery was that the 2 ${}^1\Sigma_g^+$ state has not one but two potential minima. The inner minimum had previously been called the E ${}^1\Sigma_g^+$ state and may be qualitatively described as covalent, arising from the 1s2s configuration. The outer minimum had previously been labeled a separate state, the F ${}^1\Sigma_g^+$ state, which is ionic (H^+H^-) and described by the $(2p\sigma)^2$ configuration.

[a] E. A. Hylleraas and B. Undheim, Numerische Berechnung der 2S-Terme von Ortho-und Para-Helium, *Z. Physik* **65**, 759 (1930).
[b] J. K. L. MacDonald, Successive approximations by the Rayleigh–Ritz variation method, *Phys. Rev.* **43**, 830 (1933).

23

R. E. WATSON

Approximate wave functions for atomic Be

Phys. Rev. **119**, 170 (1960)

This paper by Watson was far ahead of its time and in many respects charted the future for the self-consistent-field (SCF) configuration interaction (CI) approach.[a] As noted, there were two schools of thought as to how to proceed in electronic structure at the time. Roothaan's school at the University of Chicago was almost exclusively engaged in the quest for the Hartree-Fock limit—the very best single configuration wavefunction. Boys's school at Cambridge went well beyond the single configuration approximation, but did not consider the attainment of the near Hartree-Fock solution to be of great value in itself. Watson merged the two philosophies by taking an enormous (by 1960 standards) basis set, first performing a matrix Hartree-Fock calculation, and then proceeding to extensive CI. With six s, five p, four d, three f, and two g atomic orbitals, Watson used a 37 configuration wavefunction to achieve 89.5 per cent of the correlation energy. Watson found that doubly excited configurations (differing by two electrons from the Hartree-Fock configuration) contribute the bulk of the correlation energy, while single and triple excitations were seen to be of negligible importance. Although quadruple excitations do not interact directly with the SCF reference configuration, their interaction via the double excitations was quite strong, and four of these were included in the final CI.

[a] See I. Shavitt, The method of configuration interaction, pages 189–275 of Volume 3, *Modern theoretical chemistry*, editor H. F. Schaefer (Plenum, New York, 1977).

24

C. C. J. ROOTHAAN

Self-consistent field theory for open shells of electronic systems

Rev. Mod. Phys. **32**, 179 (1960)

As noted earlier, the unrestricted Hartree-Fock (UHF) method suffers[a] from failing to provide an exact eigenfunction of $\mathbf{S}^2$. Specifically, for open-shell singlet states, with unpaired orbitals a and b, it is apparent that a single determinant wavefunction will provide a poor description of the proper two-determinant wavefunction

$$\frac{1}{\sqrt{2}}\,[\mathrm{a}\alpha\mathrm{b}\beta - \mathrm{a}\beta\mathrm{b}\alpha]\,.$$

The restricted Hartree-Fock (RHF) theory of Roothaan extended the SCF method to properly treat such cases. For the lithium atom discussed earlier in the context of the UHF method, this means that

$$(\mathrm{RHF}) \rightarrow 1\mathrm{s}(1)\alpha(1)\;1\mathrm{s}(2)\beta(2)\;2\mathrm{s}(3)\alpha(3),$$

where the 1s spatial orbitals associated with α and β spins are now forced to be identical. This results in different Fock operators F_c and F_o for the closed and open shells and significantly complicates the resulting pseudo-eigenvalue problem. However, an exact eigenfunction of $\mathbf{S}^2$ is obtained and the variational principle satisfied, and literally thousands of molecular calculations have proved the tractability of Roothaan's RHF method.

[a] For an extreme example, see S. Bell, *Ab initio* calculations of A' and A'' states of nitrosyl cyanide, *J. Chem. Soc. Faraday Trans. 2* 77, 321 (1981).

25

B. J. RANSIL

Studies in molecular structure. I. Scope and summary of the diatomic molecule program

Rev. Mod. Phys. **32**, 239 (1960)

Ransil's paper represented the first systematic *ab initio* study of the properties of any series of molecules.[a] This work was carried out using methods developed in Roothaan's laboratory by A. Weiss, M. Yoshimine, A. D. McLean, and C. C. J. Roothaan for the UNIVAC 1103 computer. In every case reported by Ransil, a minimum basis set of Slater functions was chosen. For each diatomic molecule, however, the SCF energy was minimized with respect to every Slater function orbital exponent ζ. The greatest energy lowering obtained in this way was 0.075 51 hartree (for the N_2 ground state), relative to the energy obtained with atom-optimized orbital exponents. The set of closed shell molecules studied included LiH, BH, HF, Li_2, Be_2, C_2, N_2, F_2, CO, BF, and LiF. Dipole moments were predicted to be in fair agreement with experiment and the shocking prediction (now known to be true) made that the polarity of BF is $^{-}BF^{+}$, contrary to simple electronegativity arguments. Dissociation energies were predicted to be significantly less than experiment in every case.

[a] In recent years, of course, Pople's group has championed the systematic theoretical study of classes of molecules. See for example W. J. Hehre and J. A. Pople, Molecular orbital theory of the electronic structure of organic molecules. XXVI. Geometries, energies, and polarities of C_4 hydrocarbons, *J. Amer. Chem. Soc.* 97, 6941 (1975).

26

R. K. NESBET

Diatomic molecule project at RIAS and Boston University

Rev. Mod. Phys. **32**, 272 (1960)

This paper by Nesbet documents what is probably the first SCF wavefunction approaching the Hartree–Fock limit for a molecule larger than H_2. The paper appears in the proceedings of the 1959 American Conference on Theoretical Chemistry, held in Boulder, Colorado.[a] This was the conference at which Coulson predicted the parting of the ways between *ab initio* and semi-empirical adherents of molecular electronic structure theory.[b] As reported in the present paper, Nesbet and P. Merryman had cooperated in the development of a new diatomic molecule program, written for the IBM 704 computer. Using a basis set of roughly 'double zeta plus polarization' quality[c] (nine σ and five π Slater functions, including dσ and dπ functions), an SCF energy of −99.991 07 hartrees was obtained (Hartree-Fock limit now known to be −100.071 hartrees). Moreover, the predicted dissociation energy (4.41 eV) was in much better agreement with experiment (6.1 ± 0.2 eV) than was Ransil's minimum basis set prediction (2.56 eV).

[a] For a summary of the meeting, see K. Ruedenberg, Boulder Conference on molecular quantum mechanics, *Physics Today*, pages 34–6, May, 1960.
[b] C. A. Coulson, Present state of molecular structure calculations, *Rev. Mod. Phys.* **32**, 170 (1960).
[c] For a discussion of basis sets, see H. F. Schaefer, *The electronic structure of atoms and molecules: a survey of rigorous quantum mechanical results* (Addison-Wesley, Reading, Massachusetts, 1972).

27

S. F. BOYS & G. B. COOK

Mathematical problems in the complete quantum predictions of chemical phenomena

Rev. Mod. Phys. **32**, 285 (1960)

This is Boys's last paper on the general problem of configuration interaction. This and the following paper on CH_2 are perhaps the last of Boys's landmark papers in molecular quantum mechanics. Thus all of his most important research, which has so greatly influenced the course of molecular electronic structure theory, was published during a ten-year period. For the last decade of his life Boys's research was devoted to topics (such as transcorrelated wavefunctions[a]) which have yet to make a major impact on theoretical chemistry. In the Boys and Cook paper at hand, they lay out the general CI formulation developed at Cambridge over the previous decade. Considerable attention is paid to the problem of 'projective reduction', the term Boys gave to the determination of the coefficient of each integral appearing in each hamiltonian matrix element. Boys concluded that 'it is the projective reduction processes which always appear to have caused the most intellectual difficulty'. With the coming of powerful new methods such as the graphical unitary group approach, we can only concur with Boys in this opinion.

[a] S. F. Boys and N. C. Handy, A first solution, for LiH, of a molecular transcorrelated wave equation by means of restricted numerical integration, *Proc. Roy. Soc. (London)* **A311**, 309 (1969).

28

J. M. FOSTER & S. F. BOYS

Quantum variational calculations for a range of CH_2 configurations

Rev. Mod. Phys. **32**, 305 (1960)

The most important chemical application carried out by Boys is reported here. The study of Foster and Boys was purely predictive in nature since there was no published spectroscopic data for CH_2 when this paper was submitted in 1959. A basis set of eight Slater functions (even Boys wasn't yet entirely convinced of the efficiency of gaussians!) was used: 1s, 1s$'$, 2s, $2p_x$, $2p_y$, $2p_z$ on carbon and a single 1s function on each hydrogen atom. Although an SCF calculation was not carried out, orbitals approaching SCF quality were obtained by decreasing the importance of single excitations (to less than 10 per cent) in the CI. Large CI (for 1960) was carried out, including 128 determinantal expansion functions for the $^3B_1$ state. Thirteen different geometries were investigated, resulting in a $^3B_1$ equilibrium geometry of 129°. The $^1A_1$ state was predicted to have a bond angle of 90° and lie 24.5 kcal higher in energy. The experimental values of these parameters are now known[a,b] to be $\theta_e(^3B_1) \sim 134°$, $\theta_e(^1A_1) \sim 102.4°$, and $E(^3B_1 - {}^1A_1) \sim 9$ kcal.

[a] C. C. Hayden, D. M. Neumark, K. Shobatake, R. K. Sparks, and Y. T. Lee, Methylene singlet–triplet energy splitting by molecular beam photodissociation of ketene, *J. Chem. Phys.* **76**, 3607 (1982).
[b] P. Jensen, P. R. Bunker, and A. R. Hoy, The equilibrium geometry, potential function, and rotation–vibration energies of CH_2 in the $\tilde{X}\ ^3B_1$ ground state, *J. Chem. Phys.* **77**, 5370 (1982).

29

O. SINANOGLU

Many-electron theory of atoms and molecules

Proc. U.S. Natl. Acad. Sci. **47**, 1217 (1961)

This is one of the earliest papers describing the method which eventually came to be called 'pair correlation theory', or more precisely the independent electron pair approximation (IEPA).[a] Sinanoglu was one of the few chemists familiar with the recent developments in many-body physics (the most important contributors being Brueckner, Goldstone, and Bethe), and he coupled this knowledge with considerable physical insight. The basic idea that emerged from papers such as the present was that pair correlation energies could be calculated one at a time, and that the sum of these pair correlation energies would provide an accurate estimate of the total correlation energy. Since the calculation of each pair correlation energy was effectively a two-electron problem, for which Hylleraas techniques might be used, it is not surprising that much enthusiasm initially greated this approach. An especially interesting aspect of Sinanoglu's paper was the demonstration that the important quadruple excitations in Watson's Be wavefunctions were 'unlinked clusters' and that the coefficients of these configurations were well estimated as simple products of the coefficients of the most important doubly excited configurations.

[a] An impartial and scholarly review of the IEPA and related methods has been given by W. Kutzelnigg, Pair correlation theories, pages 129–88 of Volume 3, *Modern theoretical chemistry*, editor H. F. Schaefer (Plenum, New York, 1977).

30

I. SHAVITT & M. KARPLUS

Multicenter integrals in molecular quantum mechanics

J. Chem. Phys. **36**, 550 (1962)

In this paper Shavitt and Karplus introduced the gaussian transform method for the evaluation of multicentre integrals over Slater functions. This was the most widely used Slater integral method for non-linear polyatomic molecules during the 1960s, and is the only such method which has survived to the present day. One of the earliest applications of the gaussian transform method was to the H_3 potential surface and resulted in the classic paper by Shavitt, Stevens, Minn, and Karplus.[a] With a basis set of 1s, 1s′, and 2p functions on each hydrogen atom, SSMK were able to carry out full CI (680 configurations) for a wide range of points on the potential energy hypersurface. Much credit for the development of this general method for integrals over Slater functions should also go to Stevens,[b] who refined the method and completed the final version of the computer program. In particular, Stevens's inclusion of d function integrals by this method was an extraordinary computational achievement.

[a] I. Shavitt, R. M. Stevens, F. L. Minn, and M. Karplus, Potential-energy surface for H_3, *J. Chem. Phys.* **48**, 2700 (1968).
[b] R. M. Stevens, Accurate SCF calculation for ammonia and its inversion motion, *J. Chem. Phys.* **55**, 1725 (1971).

31

E. CLEMENTI

Correlation energy for atomic systems

J. Chem. Phys. **38**, 2248 (1963)

The development in the early 1960s of a strong research group in molecular electronic structure theory at IBM San Jose was (and continues to be) of great benefit to the field. Thinking well into the future, Clementi used the facilities at IBM to carry out systematic studies of the Hartree-Fock approximation for atomic systems. In the course of these studies Clementi provided the first quantitative values of atomic correlation energies, the latter defined by Löwdin to be the difference between the Hartree-Fock energy and the exact non-relativistic energy.[a] In the paper under discussion Clementi presented results for the atoms He through Ne, along with less precise correlation energies for Na through Ar. This work provides a firm foundation for later explicit variational studies of atomic and ultimately molecular correlation energies. Among the earliest and more important such studies was that of Veillard and Clementi,[b] who showed that the systematics of the correlation energy are somewhat simplified when the two configurations $1s^22s^22p^n$ and $1s^22p^{n+2}$ are included via the MCSCF procedure.

[a] For early estimates of atomic relativistic corrections, see H. Hartmann and E. Clementi, Relativistic correction for analytic Hartree–Fock wave functions, *Phys. Rev.* **133**, A1295 (1964).
[b] E. Clementi and A. Veillard, Correlation energy in atomic systems. IV. Degeneracy effects, *J. Chem. Phys.* **44**, 3050 (1966).

32

R. M. PITZER & W. N. LIPSCOMB

Calculation of the barrier to internal rotation in ethane

J. Chem. Phys. **39**, 1995 (1963)

Today it is difficult to imagine the herculean efforts required to complete this *ab initio* study of the ethane rotational barrier. A basis set of Slater functions was used by Pitzer and Lipscomb and all integrals precisely evaluated using the zeta-function method developed by Pitzer and Barnett[a] and the earlier discussed gaussian transform method of Shavitt and Karplus. The energies reported were not only technically correct (many early quantum mechanical studies have been, not surprisingly, found to be of limited quantitative accuracy) but gave an eminently reasonable value (3.3 kcal/mole) for the barrier to internal rotation. Subsequent studies[b] imply that there are very few problems in conformational analysis that are not well described by the Hartree-Fock level of theory. Pitzer and Lipscomb also paid considerable attention to understanding the origin of the barrier to rotation in ethane. Their work in this regard was extended some years later by Sovers, Kern, Pitzer, and Karplus,[c] who concluded from an examination of energy-optimized bond orbitals that the Pauli exclusion principle has an important influence on rotational barrier mechanisms.

[a] M. P. Barnett, The evaluation of molecular integals by the zeta function expansion, *Methods in Computational Physics* **2**, 95 (1963).

[b] For a near Hartree-Fock study of the ethane barrier see E. Clementi and H. Popkie, Analysis of the formation of the acetylene, ethylene, and ethane molecules in the Hartree-Fock model, *J. Chem. Phys.* **57**, 4870 (1972).

[c] O. J. Sovers, C. W. Kern, R. M. Pitzer, and M. Karplus, Bond function analysis of rotational barriers: ethane, *J. Chem. Phys.* **49**, 2592 (1968).

33

C. C. J. ROOTHAAN & P. S. BAGUS

Atomic self-consistent field calculations by the expansion method

Methods in Comp. Phys. **2**, 47 (1963)

Documented here is perhaps the most carefully thought out and meticulously constructed computer program in the history of electronic structure theory. This is the Roothaan-Bagus analytic atomic SCF program, written for the IBM 7030 digital computer. This program was used almost exclusively for the optimization of atomic Slater basis sets for more than a decade. Moreover, Huzinaga modified the 7090 version of the Roothaan-Bagus program (less than 300 FAP statements were changed) to accommodate gaussian functions for his classic 1965 paper. However, since the program was written entirely in IBM machine language, it ceased being of value when the last 7000 series computers began to disappear. Appreciating the importance of the code, Clementi and co-workers[a] reconstructed the Roothaan-Bagus program in

FORTRAN (a relatively machine-independent language) and added the capability to treat gaussian basis sets. The Roothaan–Bagus paper also reveals many of the computational innovations developed at Chicago and provides considerable insight into the open-shell restricted-Hartree–Fock theory for atoms.

[a] B. Roos, C. Salez, A. Veillard, and E. Clementi, A general program for calculation of atomic SCF orbitals by the expansion method, *Technical Report RJ518* (IBM Research, San Jose, August 12, 1968).

34

H. P. KELLY

Correlation effects in atoms

Phys. Rev. **131**, 684 (1963)

In 1957 Goldstone[a] completed a proof, begun by Brueckner two years earlier, of the linked-cluster perturbation expansion. Although this Brueckner–Goldstone many-body perturbation theory (MBPT) was initially designed to treat infinite nuclear matter and the infinite electron gas, in 1963 Kelly clearly demonstrated the applicability of MBPT to atoms. For finite atomic and molecular systems, the principal advantage of MBPT relative to the more traditional configuration interaction (CI) method is the former's property of 'size-consistency'. When a theoretical method is size-consistent, the total energy of two non-interacting molecules is the same whether they are treated together as one 'supermolecule' or treated separately.[b] The most straightforward application of MBPT necessitates a complete set of single-particle states (orbitals) determined by a potential V. Kelly chose V to be the Hartree–Fock potential and generated a complete set of numerical s, p, and d atomic orbitals (bound and continuum states), following the classical methods of Hartree. Kelly obtained a total correlation energy of −0.0910 hartrees, or 96.5 per cent of the Be atom correlation energy, providing considerable encouragement for the new method.

[a] J. Goldstone, Derivation of the Brueckner many-body theory, *Proc. Roy. Soc. (London)* **A239**, 267 (1957).
[b] The term size-consistent was perhaps the first used in the context of electronic structure theory by J. A. Pople, pages 51–61 of *Energy, structure, and reactivity*, editors D. W. Smith and W. B. McRae (Wiley, New York, 1973).

35

C. EDMISTON & K. RUEDENBERG

Localized atomic and molecular orbitals

Rev. Mod. Phys. **35**, 457 (1963)

In 1960 Boys suggested a mathematical procedure[a] for the determination of 'localized orbitals'. Boys sought to maximize the products of the distances between the centroids of charge of all molecular orbitals. A more satisfactory proposal, now the standard procedure for the determination of localized orbitals, was given earlier by Lennard-Jones and Pople,[b] but not implemented until this classic paper by Edmiston and Ruedenberg. Although the canonical SCF molecular orbitals are well defined (being the eigenfunctions of the Fock operator), the SCF energy is invariant to unitary transformations among the occupied orbitals. The Lennard-Jones and Pople proposal was to choose that particular unitary transformation that maximized the sum of the diagonal coulomb integrals

$$J_{ii} = \iint \phi_i^*(1)\, \phi_i^*(2) \frac{1}{r_{12}} \phi_i(1)\, \phi_i(2)\, \mathrm{d}\tau_1\, \mathrm{d}\tau_2.$$

Edmiston and Ruedenberg showed that this could be done by a series of Jacobi-like rotations. In essentially every known case, this procedure yields localized orbitals consistent with chemical intuition. Moreover, where chemical intuition is incomplete,[c] the Edmiston–Ruedenberg procedure has been of considerable qualitative value in establishing meaningful valence structures.

[a] S. F. Boys, Construction of some molecular orbitals to be approximately invariant for changes from one molecule to another, *Rev. Mod. Phys.* **32**, 296 (1960).

[b] J. E. Lennard-Jones and J. A. Pople, The molecular orbital theory of chemical valency. IV, The significance of equivalent orbitals, *Proc. Roy. Soc. (London)* **A202**, 166 (1950).

[c] For example, consider some of the boron hydrides. See D. A. Kleier, T. A. Halgren, J. H. Hall and W. N. Lipscomb, Localized molecular orbitals for polyatomic molecules. I. A comparison of the Edmiston–Ruedenberg and Boys localization methods, *J. Chem. Phys.* **61**, 3905 (1974).

36

R. K. NESBET

Computer programs for electronic wave-function calculations

Rev. Mod. Phys. **35**, 552 (1963)

As progress was made in the application of electronic structure theory, it was discovered that a serious bottleneck in CI methods arose from an unexpected source. This was the transformation of integrals over atomic basis functions (Slater functions or gaussian functions) to integrals over orthogonal molecular orbitals (usually, but not always, the SCF orbitals). The transformation in question is

$$(ij|kl) = \sum_a \sum_b \sum_c \sum_d c_{ia}\, c_{jb}\, c_{kc}\, c_{ld}\, (ab|cd),$$

where i is an orthogonal molecular orbital, a is an atomic basis function and c_{ia} the coefficient of basis function a in molecular orbital i. Since there are $O(n^4)$ of the integrals $(ij|kl)$, it would appear that $O(n^8)$ operations are required to complete the entire transformation. Should this be the case, this stage of the computation would be prohibitively time-consuming for large basis sets. It may be inferred[a] from Nesbet's paper, however, that the same transformation may be achieved by a sequence of four 'quarter transformations', requiring a total of $O(n^5)$ operations. A more straightforward presentation of these ideas was given later by Bender.[b]

[a] Nesbet actually proposes to use two successive half-transformations, requiring $0(n^6)$ operations.
[b] C. F. Bender, Integral transformations. A bottleneck in molecular quantum mechanical calculations, *J. Comput. Phys.* **9**, 547 (1972).

37

M. P. BARNETT

Mechanized molecular calculations—the POLYATOM system

Rev. Mod Phys. **35**, 571 (1963)

Visionaries have for some years imagined a futuristic 'black box' computer program, to which the bench chemist specifies a desired molecule and a series of properties of interest. After a few moments of cogitation, the computer politely returns the answers, reliable to the specified tolerances. The POLYATOM system was in a sense the first concrete step towards the above dramatized goal of chemistry by computer.[a] This first public announcement of POLYATOM suggested that either gaussian or Slater functions might eventually be used. However the addition of an advanced gaussian function capability[b] was necessary before POLYATOM became probably the most widely used such program during the remainder of the 1960s. POLYATOM was followed in succession by IBMOL,[c] MOLE,[d] and GAUSSIAN 70,[e] to cite just three of the most widely used quantum chemistry systems. Of these GAUSSIAN 70 eventually emerged as the dominant code, due to its efficiency and frequent updates, beginning with GAUSSIAN 76 and continuing on an almost annual basis.

[a] A. C. Wahl, Chemistry by computer, *Scientific American*, pages 54–70, April, 1970.

[b] I. G. Csizmadia, M. C. Harrison, J. W. Moskowitz, and B. T. Sutcliffe, Nonempirical LCAO-MO-SCF-CI calculations on organic molecules with gaussian-type functions. Part I. Introductory review and mathematical formalism, *Theoret. Chim. Acta* **6**, 191 (1966).

[c] E. Clementi and D. R. Davis, Electronic structure of large molecular systems, *J. Comput. Phys.* **1**, 223 (1966).

[d] S. Rothenberg, with P. Kollman, M. E. Schwartz, E. F. Hayes, and L. C. Allen, MOLE, a system for quantum chemistry. I. General description, *Int. J. Chem. Symp.* **3**, 715 (1970).

[e] W. J. Hehre, W. A. Lathan, R. Ditchfield, M. D. Newton, and J. A. Pople, Program No. 236, Quantum Chemistry Program Exchange, Bloomington, Indiana.

38

S. HAGSTROM & H. SHULL

The nature of the two-electron chemical bond. III. Natural orbitals for H_2

Rev. Mod. Phys. **35**, 624 (1963)

Here Hagstrom and Shull report the first conventional CI wavefunction to approach the variational energy obtained by James and Coolidge in 1933. The total energy for this 33 configuration wavefunction was −1.173 13 hartrees, slightly above the −1.173 47 obtained by the Hylleraas procedure some 30 years earlier. Perhaps more important was Hagstrom and Shull's transformation of their accurate wavefunction to natural orbital form, in which only 15 of the 33 configurations have non-vanishing coefficients. Four of the 15 configurations in the natural expansion were found to predominate, with their coefficients being 0.990 ($1\sigma_g^2$), 0.101 ($1\sigma_u^2$), 0.066 ($1\pi_u^2$), and 0.055 ($2\sigma_g^2$). The next configuration ($1\pi_g^2$) is a factor of twenty less important than the fourth. It was also noteworthy that the four configuration wavefunction based on natural orbitals recovered 96.7 per cent of the H_2 correlation energy, strengthening the idea due to Löwdin that natural orbitals provide a very rapidly convergent CI expansion. An important related paper is Davidson and Jones's natural orbital analysis[a] of the accurate Hylleraas wavefunction of Kolos and Roothaan.[b]

[a] E. R. Davidson and L. L. Jones, Natural expansion of exact wavefunctions. II. The hydrogen molecule ground state, *J. Chem. Phys.* **37**, 2966 (1962).
[b] W. Kolos and C. C. J. Roothaan, Correlated orbitals for the ground state of the hydrogen molecule, *Rev. Mod. Phys.* **32**, 205 (1960).

39

A. C. WAHL

Analytic self-consistent field wavefunctions and computed properties for homonuclear diatomic molecules

J. Chem. Phys. **41**, 2600 (1964)

The development at the University of Chicago of methods for the determination of diatomic molecule self-consistent-field wavefunctions reached its conclusion with this landmark paper. The computation of Slater basis one- and two-electron integrals was carried out using methods described in an accompanying paper.[a] The (g/u) symmetry inherent in homonuclear diatomics was exploited in an effective manner, allowing a rather close approach to the Hartree–Fock limit. A Slater basis set of size 4s 3p 1d 1f was chosen and all orbital exponents optimized at the experimental bond distance for F_2. Despite the near attainment of the Hartree–Fock limit, the energy of F_2 was found to lie 1.37 eV above that of two Hartree–Fock fluorine atoms. Thus it could be stated conclusively that electron correlation effects must be explicitly considered to provide a realistic prediction of the dissociation energy of F_2. The same qualitative result has since been found for many other diatomic (and polyatomic) molecules.[b]

[a] A. C. Wahl, P. E. Cade, and C. C. J. Roothaan, A study of two-center integrals useful in calculations on molecular structure, *J. Chem. Phys.* **41**, 2578 (1964).
[b] W. C. Ermler and R. S. Mulliken, *Diatomic molecules. Results of ab initio calculations* (Academic Press, New York, 1977).

40

M. KRAUSS

Calculation of the geometrical structure of some AH_n molecules

J. Res. Nat. Bur. Stand., Sec. A. **68**, 635 (1964)

As will be seen in several papers appearing in the next two years, we are now approaching the era in which the determination of SCF wavefunctions for non-linear polyatomic molecules becomes a prominent feature of quantum chemistry. In a certain sense Krauss's paper, appearing in November 1964, inaugurated this new age. One must remember that none of the now standard contracted gaussian basis sets were available at the time of Krauss's pioneering work. Accordingly, uncontracted gaussian basis sets were used, several of them taken from unpublished work by Huzinaga, of size typically (8s 3p) for C, N, and O and (3s) for H. Thus the quality of the basis sets was comparable or slightly higher than that of the popular 4-21G basis sets widely used today.[a] Krauss concluded 'it is evident that the approximate Gaussian calculations are adequate for the determination of the molecular geometries of first-row hydrides.' In this sense, the present paper set the stage for the single application of molecular quantum mechanics that has been most successful over the past two decades—the prediction of detailed molecular structures.

[a] P. Pulay, G. Fogarasi, F. Pang, and J. E. Boggs, Systematic *ab initio* gradient calculation of molecular geometries, force constants, and dipole moment derivatives, *J. Amer. Chem. Soc.* **101**, 2550 (1979).

41

E. CLEMENTI

Tables of atomic functions

Supplement to *IBM J. Res. Develop.* **9**, 2 (1965)

With the coming of systematic studies of classes of molecules, the need for standard basis sets for molecular calculations became apparent.

Clementi's compendium of atomic SCF calculations was perhaps the first major response to this need. For all the atoms through krypton (Z = 36) and for many positive and negative ions, near Hartree–Fock basis sets were presented. The amount of computation involved may be sensed from the fact that the Slater function orbital exponents ζ were laboriously optimized in each case. Even more vigorous minimization of the total energies with respect to these non-linear parameters was carried out by Clementi a decade later and the latter work is now the standard collection of Slater basis sets.[a] In the 1965 Tables, Clementi also presented a set of smaller optimized basis sets, labelled 'double zeta' following Richardson.[b] These basis sets, with two orbital exponents (or zetas ζ) per occupied atomic orbital, have been very popular over the years and remain a good compromise between basis set size and completeness.[c]

[a] E. Clementi and C. Roetti, Roothaan–Hartree–Fock atomic wave functions. Basis functions and their coefficients for ground and certain excited states and ionized atoms, $Z \leqslant 54$, *Atomic Data and Nuclear Data Tables* **14**, 177 (1974).

[b] J. W. Richardson, Double-ζ SCF MO calculation on the ground and some excited states of N_2, *J. Chem. Phys.* **35**, 1829 (1961).

[c] Improved double zeta Slater basis sets were reported by S. Huzinaga, A family of STO atomic basis sets, *J. Chem. Phys.* **67**, 5973 (1977).

42

S. HUZINAGA

Gaussian-type functions for polyatomic systems. I

J. Chem. Phys. **42**, 1293 (1965)

As noted earlier Krauss employed some unpublished basis sets due to Huzinaga in the former's landmark study of polyatomic first-row hydrides. Here Huzinaga presented the first systematic collection of optimized basis sets for the first-row atoms H-Ne. Although minor total energy improvements were made five years later by the more exhaustive orbital exponent optimizations of van Duijneveldt,[a] the basis sets of Huzinaga have become and remain in many laboratories a standard for the field. For the atoms B-Ne, (9s 5p) and (10s 6p) basis sets were presented, and for the nitrogen atom these sets come within

0.0056 and 0.0020 hartree, respectively, of the true Hartree–Fock energy. From these studies arose the conclusion that more than twice as many primitive gaussian functions than Slater functions are required to achieve a specified total energy. It should perhaps be noted here that the 3d orbitals of transition metal atoms (such as iron) have since been shown to require only ~50 per cent more gaussian than Slater functions for a comparable description.[b]

[a] F. B. van Duijneveldt, Gaussian basis sets for the atoms H–Ne for use in molecular calculations, *IBM Technical Research Report No. RJ945*, December 10, 1971, San Jose, California.

[b] A. J. H. Wachters, Gaussian basis set for molecular wavefunctions containing third-row atoms, *J. Chem. Phys.* **52**, 1033 (1970).

43

R. K. NESBET

Algorithm for diagonalization of large matrices

J. Chem. Phys. **43**, 311 (1965)

In the solution of several specific problems in applied mathematics, theoretical chemists have made considerably more progress than have the applied mathematicians. For example, in configuration interaction (CI) studies of atomic and molecular wavefunctions, one almost invariably desires the lowest eigenvalue and corresponding eigenvector of a large real symmetric matrix. This turns out to be a far less burdensome task than extracting all the eigenvalues and eigenvectors of the same matrix. This insight was first gained by Nesbet, and he used it to solve eigenvalue problems much larger than could be handled by conventional methods.[a] Thus Nesbet's work opened the door to much larger CIs than the 128 configuration wavefunction reported by Boys in 1960 for methylene. Moreover, the great simplicity of Nesbet's algorithm (and its descendants) was to play an important role in the development of new CI methods, specifically the direct CI approach.[b]

[a] See, for example, J. H. Wilkinson, *The algebraic eigenvalue problem* (Oxford University Press, London, 1965).

[b] B. O. Roos and P. E. M. Siegbahn, The direct configuration interaction method from molecular integrals, pages 277–318 of Volume 3, *Modern theoretical chemistry*, editor H. F. Schaefer (Plenum, New York, 1977).

44

J. M. SCHULMAN & J. W. MOSKOWITZ

Preliminary results of a self-consistent-field study of the benzene molecule

J. Chem. Phys. **43**, 3287 (1965)

After Pitzer and Lipscomb's 1963 study of the ethane molecule, the next great leap forward (in terms of size of molecule studied) was made by Schulman and Moskowitz in their study of C_6H_6. This was one of the early applications of the gaussian function version of the POLYATOM system described earlier. Two small uncontracted gaussian basis sets were used, the first of size C(3s, 1p), H(1s) and the second augmented by a set of pπ functions on each carbon atom. The latter calculation required 2.5 hours on the fastest existing computer, the IBM 7094. Contrary to the long-held notion that the π electrons lie energetically well above the σ electrons, Schulman and Moskowitz found one of the π orbitals to be more strongly binding than three of the sigma orbitals. In recent years it has of course been possible to approach the Hartree-Fock limit for benzene rather closely, and Ermler, Mulliken, and Clementi[a] have reported a benzene study using a basis set about three times larger than that of Schulman and Moscowitz. The finding that the $1a_{2u}$ orbital (a π orbital) is slightly more strongly bound than $3e_{2g}$ (a σ orbital) persists.

[a] W. C. Ermler, R. S. Mulliken, and E. Clementi, *Ab initio* SCF computations on benzene and the benzenium ion using a large contracted gaussian basis set, *J. Amer. Chem. Soc.* 98, 388 (1976).

45

P. S. BAGUS

Self-consistent-field wave functions for hole states of some Ne-like and Ar-like ions

Phys. Rev. **139**, A619 (1965)

A single-configuration Hartree–Fock wavefunction provides an energetic upper bound for the lowest electronic state of its particular symmetry. For the argon atom this means that the positive ion states corresponding to the electron configurations $1s^2\,2s^2\,2p^6\,3s^2\,3p^5\,(^2P)$ and $1s^2\,2s^2\,2p^6\,3s\,3p^6\,(^2S)$ are readily accessible via Hartree–Fock theory. However the ^2S state arising from the removal of a 1s electron is by no means the lowest of its symmetry, and it was not apparent at the time of this paper that Hartree–Fock theory would be applicable. Bagus showed that such states are in fact reasonably described by SCF theory, without an explicit requirement of orthogonality to energetically lower-lying SCF states. The orbital or shell structure of the atom is sufficient to guarantee physically reasonable results. Furthermore, considerable improvement over the Koopmans' theorem[a] results (ionization potential = orbital energy) is found from such direct hole state calculations for Ne and Ar. This paper provides the theoretical framework for much of the later interpretation of the new field of photoelectron spectroscopy.[b]

[a] T. Koopmans, Über die Zuordnung von Wellenfunktionen und Eigenwerten zu den Einzelnen Elektronen eines Atoms, *Physica* **1**, 104 (1933).

[b] For early theoretical work on multiplet splittings in X-ray photoelectron spectroscopy, see P. S. Bagus and H. F. Schaefer, Direct near-Hartree–Fock calculations on the 1s hole states of NO^+, *J. Chem. Phys.* **55**, 1474 (1971).

46

G. DAS & A. C. WAHL

Extended Hartree–Fock wavefunctions: optimized valence configurations for H_2 and Li_2, optimized double configurations for F_2

J. Chem. Phys. **44**, 87 (1966)

The importance of this study by Das and Wahl was immediately recognized and a plausible case may be made that it was the most important theoretical electronic structure paper to appear during the 1960s. It includes the first applications of the multi-configuration self-consistent-field (MCSCF) method to molecular systems. The method used by Das and Wahl was an extension of the single configuration Hartree–Fock–Roothaan approach to include up to four configurations in a completely variational manner. For Li_2, to the Hartree–Fock configuration $1\sigma_g^2\ 1\sigma_u^2\ 2\sigma_g^2$ were added terms to describe left-right correlation ($2\sigma_g^2 \rightarrow 2\sigma_u^2$), angular correlation ($2\sigma_g^2 \rightarrow 1\pi_u^2$), and in-out (or radial) correlation ($2\sigma_g^2 \rightarrow 3\sigma_g^2$).[a] The particular chemical problem which instigated this study was the failure of the Hartree–Fock method to yield an F_2 energy below that of two separated Hartree–Fock fluorine atoms. The two-configuration (TC) SCF wave function of Das and Wahl did provide proper dissociation to F + F, whereas the one-configuration restricted SCF function for F_2 dissociates to an admixture of F + F and $F^+ + F^-$ fragments. Later, larger MCSCF studies of F_2 by the same authors[b] provided a quantitatively accurate prediction of the F_2 dissociation energy.

[a] These terms appear to have acquired general usage by 1966. Their first explicit enumeration was perhaps given by E. Callen, Configuration interaction applied to the hydrogen molecule, *J. Chem. Phys.* **23**, 360 (1955).

[b] G. Das and A. C. Wahl, Theoretical study of the F_2 molecule using the method of optimized valence configurations, *J. Chem. Phys.* **56**, 3532 (1973).

47

P. F. FOUGERE & R. K. NESBET

Electronic structure of C_2

J. Chem. Phys. **44**, 285 (1966)

Prior to this time nearly all *ab initio* studies of diatomic molecules had concentrated primarily or exclusively on the electronic ground state. A notable exception was Meckler's early gaussian basis study[a] of the oxygen molecule. In the present paper, Fougere and Nesbet set the stage for many future studies by carrying out a high-level study of the potential energy curves for no less than twenty-seven different electronic states of C_2. A double zeta plus polarization (DZ + P) basis set of Slater functions, C(4s 2p 1d), was used, and a single set of molecular orbitals chosen to describe all 27 electronic states. These molecular orbitals were obtained from a matrix SCF calculation on the hypothetical $^5\Delta_g$ state arising from the electron configuration $1\sigma_g^2\ 1\sigma_u^2\ 2\sigma_g^2\ 2\sigma_u^2\ 3\sigma_g\ 1\pi_u\ 1\pi_g\ 3\sigma_u$, in which all of the valence orbitals are occupied. This guarantees that these eight orbitals will change smoothly from their molecular C_2 character to the SCF orbitals of two separated carbon atoms. Using the resulting 'minimum basis' set of MOs, Nesbet carried out a uniform level of CI (up to 63 configurations) for each state, and obtained consistently reasonable agreement with experiment.

[a] A. Meckler, Electronic energy levels for molecular oxygen, *J. Chem. Phys.* **21**, 1750 (1953).

48

J. L. WHITTEN

Gaussian lobe function expansions of Hartree–Fock solutions for the first-row atoms and ethylene

J. Chem. Phys. **44**, 359 (1966)

For gaussian basis sets to become widely applicable, some mechanism had to evolve to deal with the fact that a relatively large (compared to Slater functions) number of these functions were required to reasonably approach the atomic Hartree–Fock energies. The solution to this problem was provided by Whitten, and the concept of 'contracted' gaussian functions was perhaps first introduced in this paper.[a] Since two-electron integrals over gaussian functions were relatively easy to evaluate, it was proposed to use fixed linear combinations of gaussian functions as a basis function. For the carbon atom, Whitten specifically endorsed the use of three groups of s functions, composed respectively of 3, 4 and 3 primitive gaussians, representing short, intermediate, and long-range functions. The atomic 2p orbital was represented as a single linear combination of five primitive gaussians. For more exhaustive studies, an 'extended' basis was proposed in which the most diffuse primitive s function and the most diffuse primitive p_x, p_y, p_z functions were given complete freedom with respect to their linear coefficients in each molecular orbital. A sample calculation on ethylene with this double zeta-like basis resulted in the lowest energy reported prior to 1966 for that molecule.

[a] The term 'contracted' gaussian function was apparently introduced by E. Clementi and D. R. Davis, Electronic structure of large molecular systems, *J. Comput. Phys.* **1**, 223 (1966). Clementi and Davis were certainly among the first to apply contracted gaussians to molecular systems.

49

H. J. SILVERSTONE & O. SINANOGLU

Many-electron theory of nonclosed-shell atoms and molecules. I. Orbital wavefunction and perturbation theory

J. Chem. Phys. **44**, 1899 (1966)

The papers of Hartree, Hartree, and Swirles (1939), Boys (1950), and Watson (1960) led to the conclusion that for the first row atoms Be–O, the configuration $2s^2 \rightarrow 2p^2$ (when allowed by symmetry) will be second in importance only to the Hartree–Fock configuration

$1s^2 2s^2 2p^n$. Here Silverstone and Sinanoglu generalized this observation in an important way. They distinguished between three types of doubly excited configurations: (a) internal correlation, or excitation into the unfilled region of the Hartree-Fock sea (the 1s, 2s, $2p_x$, $2p_y$, $2p_z$ orbitals for B–Ne). The $2s^2 \rightarrow 2p^2$ excitation in beryllium, although not recognized here because Be was artificially required to be considered separately as a closed-shell system, is the classic example of this sort of internal correlation effect; (b) semi-internal correlation, such as $2s^2 \rightarrow 2p3p$ for the beryllium atom, where one (2p) of the orbitals excited to lies within the Hartree–Fock sea and one (3p) lies above the Fermi level; (c) external correlations, such as $2s^2 \rightarrow 3p^2$ in which both electrons move into an orbital (or two orbitals) above the Fermi level. This categorization of configurations and its subsequent generalizations by other workers have proved to be of great value to electronic structure theory.[a]

[a] For some early calculations on atomic systems, see O. Sinanoglu and I. Oksüz, Theory of atomic structure including electron correlation, *Phys. Rev. Letters* **21**, 507 (1968).

50

P. E. CADE, K. D. SALES, & A. C. WAHL

Electronic structure of diatomic molecules. III. A. Hartree–Fock wavefunctions and energy quantities for $N_2(X\ ^1\Sigma_g^+)$ and $N_2^+(X\ ^2\Sigma_g^+, A\ ^2\Pi_u, B\ ^2\Sigma_u^+)$ molecular ions

J. Chem. Phys. **44**, 1973 (1966)

At least two procedures for the optimization of molecular basis sets were in the early 1960s thought to be reasonable. The first was to begin with a small, perhaps minimum (1s, 2s, 2p for nitrogen), basis set, optimize this exhaustively, and then add and successively optimize additional basis functions until saturation is achieved. Here Cade, Sales, and Wahl demonstrated clearly for N_2 that a much more practical and efficient approach is to begin with a large optimized atomic basis set, add suitable polarization functions (in this case d and f functions on each nitrogen atom) and carry out vigorous orbital exponent optimization. More recent studies have shown that with experience as a guide one can usually omit the final step of exponent optimization with little loss in the calculated total energy. Cade, Sales, and Wahl also reported SCF wavefunctions of comparable quality for the three lowest electronic states of the N_2^+ positive ion. Although the $^2\Sigma_g^+$ ground state is known to lie 9167 cm^{-1} below the $^2\Pi_u$ excited state, Hartree–Fock theory reverses this order and predicts the $^2\Pi_u$ state to lie 5290 cm^{-1} lower.[a] This qualitative failure of the SCF method added further impetus to the development of post-Hartree–Fock methods.

[a] This discrepancy with experiment is lessened when one appreciates that the $^2\Sigma_g^+$ RHF wavefunction exhibits symmetry breaking with respect to the inversion operator. See S. C. de Castro, H. F. Schaefer, and R. M. Pitzer, Electronic structure of the N_4^+ molecular ion, *J. Chem. Phys.* **74**, 550 (1981).

51

S. D. PEYERIMHOFF, R. J. BUENKER, & L. C. ALLEN

Geometry of molecules. I. Wavefunctions for some six- and eight-electron polyhydrides

J. Chem. Phys. **45**, 734 (1966)

The basis sets and methods developed by Whitten were very quickly put to effective use by Allen's group at Princeton University. Recognizing the great importance of Walsh's rules as a simple model for the interpretation of molecular geometries, Peyerimhoff, Buenker, and Allen inaugurated a series of papers designed to explore the relationship between Walsh's rules and *ab initio* concepts. As noted in 1963 by Coulson and Neilson[a] the sum of Hartree–Fock orbital energies may be sufficiently different from the SCF total energy that the orbital energies form a poor approximation to the 'orbital binding energies' discussed by Walsh in his classic 1953 papers. Using Whitten's DZ-like basis sets, the authors showed that for the most part the SCF orbital energy curves (as a function of HAH bond angle) have the same shapes as the orbital binding energy curves presented by Walsh thirteen years earlier. In this regard Davidson[b] has more recently provided a definition of orbital energies in which the sum of the latter *is* equal to the total energy, thus providing a firm theoretical foundation for Walsh's rules.

[a] C. A. Coulson and A. H. Neilson, Angular correlation diagrams for AH_2-type molecules, *Discussions Faraday Soc.* **35**, 71 (1963).
[b] L. Stenkamp and E. R. Davidson, An internally consistent SCF investigation of Walsh's rules, *Theoret. Chim. Acta* **30**, 283 (1973).

52

C. EDMISTON & M. KRAUSS

Pseudonatural orbitals as a basis for the superposition of configurations. I. He_2^+

J. Chem. Phys. **45**, 1833 (1966).

For two-electron systems, the transformation to natural orbitals reduces the length of the CI expansion from $O(n^2)$ to $O(n)$ terms. Accordingly, a number of theoreticians in 1966 were pondering how to make comparable use of natural orbitals for many (i.e., >2) electron systems. Perhaps the first successful proposal was that of Edmiston and Krauss, who introduced the notion of pseudonatural orbitals. This method is of greatest value when all the orbitals being correlated lie in the same general region of space. Thus it would not be optimally effective for a system such as LiH with one core and a second physically removed valence orbital. For He_2^+, with electron configuration $1\sigma_g^2\ 1\sigma_u$, the two occupied SCF orbitals are spatially similar, and Edmiston and Krauss obtained NOs for the $1\sigma_g^2$ pair of electrons in the field of a singly-occupied $1\sigma_u$ molecular orbital. These pseudonatural orbitals were subsequently used in CI wavefunctions in which correlation effects for all three electrons were described. It is a relatively simple matter to extend the Edmiston–Krauss concept. In perhaps the first such extrapolation,[a] NOs obtained by correlating the three highest occupied SCF MOs of C_3 were used in a subsequent CI treatment of all twelve valence electrons.

[a] D. H. Liskow, C. F. Bender, and H. F. Schaefer, Bending frequency of the C_3 molecule, *J. Chem. Phys.* **56**, 5075 (1972).

53

J. CIZEK

On the correlation problem in atomic and molecular systems. Calculation of wavefunction components in Ursell-type expansion using quantum-field theoretical methods

J. Chem. Phys. **45**, 4256 (1966)

Among non-variational alternatives to the CI method, the coupled cluster (CC) method is certainly one of the most promising. There seems to be general agreement that much work on this topic in chemical physics may be traced to Cizek's 1966 paper. Cizek's paper is in turn indebted to earlier work by the nuclear physicists Coester and Kümmel,[a] who introduced the exponential ansatz

$$\psi = e^{T} \mid \Phi_o \rangle,$$

in which Φ_o is a suitable reference function, typically the Hartree-Fock wavefunction. T is an excitation operator

$$T = T_1 + T_2 + \ldots T_n,$$

for which the subscripts refer to the number of electrons excited in Φ_o. Cizek presented this theory in a form suitable for molecular quantum mechanics and carried out the first electronic applications[b] of the formalism. The most commonly applied form of the theory is the coupled-cluster doubles (CCD) method, in which $T = T_2$. The terms T_1 and T_3 first appear in the fourth order of perturbation theory and T_4 in fifth order, providing the justification for the frequent neglect of these effects.

[a] F. Coester and H. Kümmel, Short range correlations in nuclear wave functions, *Nucl. Phys.* **17**, 477 (1960).
[b] The first *ab initio* study of the coupled cluster method was the minimal basis study of J. Paldus, J. Cizek, and I. Shavitt, Correlation problems in atomic and molecular system. IV. Extended coupled-pair many-electron theory and its application to the borane molecule, *Phys. Rev. A* **5**, 50 (1972).

54

C. F. BENDER & E. R. DAVIDSON

A natural orbital based energy calculation for helium hydride and lithium hydride

J. Phys. Chem. **70**, 2675 (1966)

Perhaps the most important formulation of natural orbital concepts since their introduction by Löwdin in 1955 resulted in the iterative natural orbital (INO) method. Suppose one wishes to obtain approximate natural orbitals for a molecular system where the number of potentially important configurations is vastly in excess of the number which can be treated variationally. Such was the difficulty facing Bender and Davidson for LiH with a large basis set (17σ, 10π, 5δ) and a limitation of 50 configurations. Beginning with the occupied SCF orbitals and a virtual set chosen to diagonalize the SCF exchange integral operator, the authors chose the 50 configurations thought to be most important and carried out a CI. From the CI vector the density matrix was constructed and diagonalized to give a set of approximate natural orbitals. Using these NOs a new set of 50 configurations (~35 of which had been used previously) was chosen and the CI repeated to determine a second set of approximate NOs. When the wavefunction and orbitals converged, Bender and Davidson had a 45-configuration wavefunction including 88.7 per cent of the correlation energy, a result which required nearly a decade to be improved upon.[a]

[a] W. Meyer and P. Rosmus, PNO–CI and CEPA studies of electron correlation. III. Spectroscopic constants and dipole moment functions for the ground states of the first-row and second-row diatomic hydrides, *J. Chem. Phys.* **63**, 2356 (1975). These authors obtained 93.6 per cent of the LiH correlation energy in their variational treatment.

55

A. D. McLEAN & M. YOSHIMINE

Tables of linear molecule wave functions

Supplement to *IBM J. Res. Develop.* **12**, 206 (1968)

By 1967 the group at the University of Chicago had determined near-Hartree–Fock wavefunctions for essentially all first-row homonuclear and heteronuclear diatomic molecules. However, several years passed before these results were made generally available.[a] At roughly the same time that Wahl, Cade, and Roothaan were putting the finishing touches on their diatomic SCF programs, at IBM San Jose, McLean and Yoshimine were developing a powerful new linear molecule program system for the IBM 7094. The fruits of this work appeared in the widely circulated 'Tables of linear molecule wave functions', which included SCF studies of systems as large as cyanoacetylene, N≡C–C≡CH. The smallest basis sets used were of double zeta plus polarization (DZ + P) quality, that is, very respectable. For some of the smaller molecules studied, such as HCN, much larger basis sets were used and several points on the potential energy hypersurface were considered. As the first collection of high-level SCF wavefunctions for polyatomic molecules, the McLean–Yoshimine report was also of great value to theorists studying the qualitative aspects of chemical bonding.[b]

[a] P. E. Cade and W. M. Huo, Hartree–Fock wavefunctions for diatomic molecules. III. First row heteronuclear systems, AB, $AB^{\pm}$, and AB^{*}, *Atomic Data Nuclear Data Tables* **15**, 1 (1975).

[b] For example, an illuminating colour representation of the total electron density of fluoroacetylene was created by W. E. Donath and appears on the cover of the May, 1968 issue of the *IBM Journal of Research and Development.*

56

F. GRIMALDI, A. LECOURT, & C. MOSER

The calculation of the electric dipole moment of CO

Int. J. Quantum Chem. Symp. **1**, 153 (1967)

Robert Mulliken once stated something to the effect that when the value of a particular molecular property is very small in absolute terms, quantum mechanics ought to be able to predict the magnitude of that quantity, but not necessarily the correct sign. In her careful 1965 SCF study of the CO dipole moment, Huo[a] predicted a value of 0.15 debye (0.27 debye at the experimental equilibrium bond distance), with the intuitively expected polarity C^+O^-. Although the magnitude of this prediction was correct, the experimental polarity was an unexpected C^-O^+. The following paper (in this commentary) by Yoshimine and McLean used a much larger basis set and found a near-Hartree-Fock dipole moment (at the experimental equilibrium bond distance) of 0.28 debye, again with the incorrect polarity. In the present paper Grimaldi, Lecourt, and Moser showed that a CI of 200 singly and doubly excited configurations changes the SCF value of 0.15 debye (C^+O^-) by 0.34 debye to 0.19 debye (C^-O^+). Strikingly the final agreement with experiment is not found until the singly excited configurations (quite unimportant in determining the correlation energy) are included in the CI.[b]

[a] W. M. Huo, Electronic structure of CO and BF, *J. Chem. Phys.* **43**, 624 (1965).

[b] This turns out to be true in general. See S. Green, Sources of error and expected accuracy in *ab initio* one-electron operator properties: the molecular dipole moment, *Adv. Chem. Phys.* **25**, 179 (1974).

57

M. YOSHIMINE & A. D. McLEAN

Ground states of linear molecules: dissociation energies and dipole moments in the Hartree–Fock approximation

Int. J. Quantum Chem. Symp. **1**, 313 (1967)

Here Yoshimine and McLean extracted some of the most important observables from their previously cited 'Tables of linear molecule wave functions' and compared them with available experimental data. They also documented the energetic effect of adding various types of polarization basis functions to the N_2O molecule. As a general rule, predicted near-Hartree–Fock dissociation energies were found to be notably less than experiment. The CO_2 molecule is fairly typical in this regard, the near-Hartree–Fock dissociation energy (to $C + O + O$) being 11.3 eV, as opposed to the experimental value 16.9 eV. Following the definition of Clementi,[a] the 'molecular extra correlation energy', that part of the molecular correlation energy not accounted for by the sum of the separated atom correlation energies, is $16.9 - 11.3 = 5.6$ eV for CO_2. Dipole moments were generally found to be in good agreement with experiment, the theoretical values being typically in error by ~0.2 debye. Ransil's 1960 minimum basis set SCF prediction that the polarity of BF is $^{-}BF^{+}$ was confirmed at a much higher level of the theory.

[a] E. Clementi, SCF-MO wave functions for the hydrogen fluoride molecule, *J. Chem. Phys.* **36**, 33 (1962).

58

A. A. FROST

Floating spherical Gaussian orbital model of molecular structure I. Computational procedure. LiH as an example

J. Chem. Phys. **47**, 3707 (1967)

What is the simplest imaginable, all-electron, *ab initio* approach to molecular electronic structure? Within an orbital framework (in this writer's view, chemistry has come too far on the shoulders of the orbital perspective during the past 50 years[a] to abandon this picture now), each pair of electrons might be described by a single basis function. For added simplicity that basis function might be taken as a spherical gaussian function (i.e., 1s gaussian basis function). It follows that in such a model there is no need for a self-consistent-field procedure since the number of electron pairs is identical to the number of basis functions. This is precisely the model suggested and implemented by Frost in his 1967 paper. Frost further required in his FSGO model that the total energy be minimized with respect to position and orbital exponent of each spherical gaussian. The conceptual beauty of the model is apparent and applications have been widespread. Christoffersen subsequently developed a more elaborate method[b] in which preoptimized FSGO fragments are brought together to provide a theoretical treatment of larger molecules.

[a] R. Hoffmann, Building bridges between inorganic and organic chemistry, *Agnew. Chem. Int. Ed. Engl.* **21**, 711 (1982).
[b] R. E. Christoffersen, D. W. Genson, and G. M. Maggiora, *Ab initio* calculations on large molecules using molecular fragments. Hydrocarbon characterizations. *J. Chem. Phys.* **54**, 239 (1971).

59

E. CLEMENTI & J. N. GAYLES

Study of the electronic structure of molecules. VII. Inner and outer complex in the NH_4Cl formation from NH_3 and HCl

J. Chem. Phys. **47**, 3837 (1967)

Over the past fifteen years there have been a fair number of molecules whose existence was suggested by *ab initio* theory prior to experimental observation.[a] Perhaps the first such example was Clementi's work on the gas phase ammonium chloride molecule. Using an early version of what was to become the IBMOL program, Clementi carried out a reasonably complete search of the NH_4Cl potential energy hypersurface in the vicinity of suspected potential minima. The SCF approximation was used in conjunction with a basis set of roughly double zeta calibre. In this study it was concluded that NH_4Cl is best described as $H_3N\cdots HCl$ rather than $NH_4^+\cdots Cl^-$. This paper was also among the earliest to attempt a theoretical prediction of the vibrational frequencies and thermodynamic properties of a polyatomic molecule. The laboratory observation of NH_4Cl in 1969 represented a major triumph for theoretical chemistry.[b,c]

[a] A more recent example is the $Li\cdots H_2O$ complex, predicted by theory to be bound by ~12 kcal. See M. Trenary, H. F. Schaefer, and P. A. Kollman, Electronic structure of $Li–H_2O$ and related neutral molecular complexes, including $Al–H_2O$, *J. Chem. Phys.* **68**, 4047 (1978).

[b] P. Goldfinger and G. Verhaegen, Stability of the gaseous ammonium chloride molecule, *J. Chem. Phys.* **50**, 1467 (1969).

[c] For a more recent study of the same system, see R. C. Raffenetti and D. H. Phillips, Gaseous NH_4Cl revisited: a computational investigation of the potential surface and properties, *J. Chem. Phys.* **71**, 4534 (1979).

60

R. K. NESBET

Atomic Bethe–Goldstone equations. I. The Be atom

Phys. Rev. **155**, 51 (1967)

The first consistently high level *ab initio* application of the independent electron pair approximation (IEPA) was reported in this paper. Nesbet presented his results in the framework of a model which for the closed-shell beryllium atom allows one to distinguish between two-, three, and four-electron correlation effects. The sum of the pair correlation energies amounted to −0.093 04 hartree, or an impressive 98.7 per cent of the experimental correlation energy. Nesbet's correction (unique to his formalism) for triple excitations was obtained using a smaller (but still reasonable) basis set and was nearly negligible, +0.000 92 hartree. Thus the final predicted correlation energy was −0.092 13 hartree, or 97.7 per cent of the observed correlation energy. Nesbet carried out comparable studies of all the other first-row atoms[a] and obtained equally good agreement with experiment. In retrospect it seems clear that the euphoria[b] associated with the IEPA reached a peak with the response to the publication of Nesbet's landmark papers.

[a] R. K. Nesbet, Atomic Bethe–Goldstone equations. III. Correlation energies of Be, B, C, N, O, F, and Ne, *Phys. Rev.* **175**, 2 (1968).
[b] O. Sinanoglu, Electron correlation in atoms and molecules, *Adv. Chem. Phys.* **14**, 237 (1969).

61

R. AHLRICHS & W. KUTZELNIGG

Direct calculation of approximate natural orbitals and natural expansion coefficients of atomic and molecular electronic wavefunctions. II. Decoupling of the pair equations and calculation of the pair correlation energies for the Be and LiH ground states

J. Chem. Phys. **48**, 1819 (1968)

Many papers, beginning with those of Sinanoglu,[a] were devoted in the 1960s to the theoretical justification of the IEPA. As is often the case, the passage of several years time helped to clarify some of this thinking, and the Ahlrichs-Kutzelnigg paper gives a lucid picture of IEPA's theoretical basis as of 1968. Unique to the work in Kutzelnigg's group was the conviction that each 'decoupled pair equation' should be solved in terms of the natural orbitals appropriate to that particular electron pair. This concept was later exploited remarkably effectively by Meyer in his development of the variational pair natural orbital (PNO) CI method. In one of the earliest applications of the IEPA to a molecular system,[b] Ahlrichs and Kutzelnigg reported results for lithium hydride as noted in the title. Using a gaussian basis set, about 78 per cent of the LiH correlation energy was obtained. Most of the 22 per cent of the correlation energy not obtained appeared to reside in the $1\sigma^2$ core electron pair, for which the chosen gaussian basis set was not particularly suitable.

[a] O. Sinanoglu, Many-electron theory of atoms and molecules. I. Shells, electron pairs versus many-electron correlation, *J. Chem. Phys.* **36**, 706 (1962).

[b] For a later IEPA study by the Kutzelnigg group, see M. Jungen and R. Ahlrichs, *Ab initio* calculations on small hydrides including electron correlation. III. A study of the valence shell intrapair and interpair correlation energy of some first row hydrides, *Theoret. Chim. Acta* **17**, 339 (1970).

62

K. MOROKUMA & L. PEDERSEN

Molecular-orbital studies of hydrogen bonds. An *ab initio* calculation for dimeric H_2O

J. Chem. Phys. **48**, 3275 (1968)

Clementi's early studies of $H_3N \cdots HCl$ in fact represented the first *ab initio* investigation of the hydrogen bond. However, that was something of a surprise to those who expected NH_4Cl to have the ionic structure $NH_4^+Cl^-$. The first study of a 'traditional' hydrogen bond was that of Morokuma and Pedersen, then postdoctoral fellows with Martin Karplus, on the water dimer. A small uncontracted gaussian basis O(5s 3p), H(3s) was used in conjunction with the SCF approximation. A dimerization energy of 12.6 kcal was predicted, about twice as large as estimated experimentally. In retrospect one now knows that basis sets of this type (roughly double zeta in calibre) without polarization functions (d functions on oxygen; p functions on hydrogen) uniformly overestimate the strength of hydrogen bonds.[a] Interestingly, Clementi has shown[b] that as one goes beyond DZ + P basis sets and closely approaches the Hartree-Fock limit, the hydrogen bond energy becomes *less than* the experimental estimates. Clementi's largest basis set predicts a stabilization energy of only 3.9 kcal for the water dimer within the SCF approximation.

[a] P. A. Kollman, Hydrogen bonding and donor-acceptor interactions, pages 109–52 of Volume 4, *Modern theoretical chemistry*, editor H. F. Schaefer (Plenum, New York, 1977).

[b] H. Popkie, H. Kistenmacher, and E. Clementi, Study of the structure of molecular complexes. IV. The Hartree–Fock potential for the water dimer and its application to the liquid state, *J. Chem. Phys.* **59**, 1325 (1973).

63

W. KOLOS & L. WOLNIEWICZ

Improved theoretical ground-state energy of the hydrogen molecule

J. Chem. Phys. **49**, 404 (1968)

Beginning in 1960,[a] Kolos and his colleagues (initially Roothaan, later Wolniewicz) reported a series of increasingly sophisticated, nearly exact solutions to the Schrödinger equation for the different isotopic variants of the hydrogen molecule. Their wavefunctions were explicitly dependent on the interelectronic distance r_{12} and thus may be viewed as an extension of the earlier-discussed work of James and Coolidge. The lowest Born–Oppenheimer (clamped nucleus) energy obtained in the present paper was −1.174 475 hartrees (or 0.001 hartree below the 1933 James and Coolidge result) and may be taken to be exact to the seven significant figures quoted. Moreover, the papers in the Kolos series went well beyond the Born–Oppenheimer picture to treat the electronic and vibrational degrees of freedom simultaneously. Furthermore their results were subsequently corrected for both relativistic and radiative effects. In this manner the dissociation energy D_0(H–H) was predicted to be 36 117.4 cm^{-1}, outside the error bars of the experimental value $D_0 = 36\ 113.6 \pm 0.3\ cm^{-1}$. This discrepancy was resolved one year later when Herzberg's new analysis[b] of the spectroscopic data yielded the definitive experimental result $D_0(H_2) = 36\ 117.3 \pm 1.0\ cm^{-1}$.

[a] W. Kolos and C. C. J. Roothaan, Accurate electronic wave functions for the H_2 molecule, *Rev. Mod. Phys.* **32**, 219 (1960).

[b] G. Herzberg, Dissociation energy and ionization potential of molecular hydrogen, *Phys. Rev. Letters* **23**, 1081 (1969).

64

E. R. DAVIDSON & C. F. BENDER

Correlation energy calculations and unitary transformations for LiH

J. Chem. Phys. **49**, 465 (1968)

Bender and Davidson's communication provided the first of an eventually crippling series of blows to the IEPA. There are two obvious choices of molecular orbitals for LiH—the canonical SCF orbitals and the Edmiston-Ruedenberg localized orbitals (the two sets are actually rather similar for LiH). It had been implicitly assumed by many that the sum of the pair correlation energies would be very similar for these two choices of molecular orbitals. Bender and Davidson calculated pair correlation energies for enough orthogonal possibilities for the reference function orbitals 1σ and 2σ to span the entire range of possible choices, and their results were presented in a helpful pictorial form. The CI total energy including all double excitations is invariant to the unitary transformation from the canonical SCF orbitals 1σ and 2σ. However, it was found that there is a broad range of orbital choices for which the sum of the pair correlation energies is very different (by ~25 per cent) from the estimated full CI energy. An analogous study of the BH molecule was reported the following year.[a]

a C. F. Bender and E. R. Davidson, Unitary transformations and pair energies, *Chem. Phys. Lett.* **3**, 33 (1969).

65

J. GERRATT & I. M. MILLS

Force constants and dipole-moment derivatives of molecules from perturbed Hartree–Fock calculations. I

J. Chem. Phys. **49**, 1719 (1968)

It goes almost without saying that certain scientific papers while recognized as important upon publication, become much more important as time passes. This is certainly true of the paper by Gerratt and Mills, who set out to determine the derivatives of each occupied SCF molecular orbital with respect to each nuclear coordinate. As methods have developed since 1978 for the determination of analytic SCF energy second derivatives (the use envisaged by the authors in 1968) and analytic CI energy first derivatives, the solution of these 'coupled-perturbed Hartree–Fock equations' has become absolutely essential. Detailed studies of potential energy hypersurfaces are tremendously simplified by these powerful new techniques. Accordingly, the Gerratt and Mills paper has become a genuine classic in the history of molecular electronic structure theory. It should be noted that related, although somewhat simpler, equations were presented[a] and solved five years earlier for the changes in SCF molecular orbitals with respect (not to nuclear coordinates but) to an external electric or magnetic field.

[a] R. M. Stevens, R. M. Pitzer, and W. N. Lipscomb, Perturbed Hartree–Fock calculations. I. Magnetic susceptibility and shielding in the LiH molecule, *J. Chem. Phys.* **38**, 550 (1963).

66

D. NEUMANN & J. W. MOSKOWITZ

One-electron properties of near-Hartree–Fock wave functions; I. Water

J. Chem. Phys. **49**, 2056 (1968)

As SCF wavefunctions became generally available for non-linear polyatomic molecules, there was rather naturally a concomitant demand for theoretical predictions of a variety of observables. Neumann and Moskowitz provided an important service to the theoretical community by incorporating methods for the determination of a large number of one-electron properties into the POLYATOM system of programs. The first application of these methods is seen in the present paper on the water molecule. The single-configuration SCF approximation was used in conjunction with a large uncontracted gaussian basis set, designated O(10s 6p 2d) H(4s 2p). An important complementary paper on water, but using a Slater function basis set of size O(3s 2p 1d), H2s), was published simultaneously by Pitzer and co-workers.[a] The two sets of theoretical predictions were generally in good agreement with each other and with available experimental data. For example, the dipole moments were 2.04 debye (Pitzer), 2.00 debye (Neumann-Moskowitz), and 1.85 ± 0.02 debye (experiment). The methods developed by Neumann and Moskowitz have since been exploited to predict important properties not considered by NM, such as magnetic hyperfine parameters.[b]

[a] S. Aung, R. M. Pitzer, and S. I. Chan, Approximate Hartree–Fock wavefunctions, one-electron properties, and electronic structure of the water molecule, *J. Chem. Phys.* **49**, 2071 (1968).

[b] H. F. Schaefer and S. Rothenberg, Magnetic hyperfine structure of NO_2, *J. Chem. Phys.* **54**, 1432 (1971).

67

C. F. BUNGE

Electronic wave functions for atoms. I. Ground state of Be

Phys. Rev. **168**, 92 (1968)

As noted earlier, Watson's 37 configuration wavefunction, employing a 9s 7p 5d 3f 3g Slater basis set, achieved 89.5 per cent of the Be correlation energy in 1960. Bunge took precisely the same basis set, omitted the f and g basis functions completely, and then proceded to variationally recover 97.1 per cent of the correlation energy. How did he do it? First, but not most important, Bunge used a larger CI, a total of 180 terms. To within 0.0005 hartree, however, the same total energy is obtained with only 91 configurations. More crucially, Bunge made very creative uses of the concept of natural orbitals. Pair natural orbitals were initially investigated for all three orbital pairs, $1s^2$, 1s2s, and $2s^2$, and it was observed that the individual 1s2s NOs resembled either the $1s^2$ NOs or the $2s^2$ NOs. Therefore $1s^2$ and $2s^2$ NOs were separately determined and the most important of each merged and orthogonalized to form a single set of atomic orbitals. Among other innovations Bunge presented a scheme for the determination of variational pair correlation energies, such that their sum is precisely equal to the correlation energy obtained via CI. The exact values of these pair energies were estimated to be $\epsilon(1s^2) = 0.0426$, $\epsilon(2s^2) = 0.0455$, and $\epsilon(1s,2s) = 0.0053$ hartree.

68

H. F. SCHAEFER & F. E. HARRIS

Metastability of the ^1D state of the nitrogen negative ion

Phys. Rev. Lett. **21**, 1561 (1968) (First-order wave function.)

A variational generalization of the earlier discussed work of Silverstone and Sinanoglu was provided here in the form of the 'first-order' wavefunction. In many ways this name is unfortunate, but it appears to have stood the test of time in theoretical circles. The thought intended by the designation first order was inspired by the theorem of Møller and Plesset, that closed-shell Hartree–Fock wavefunctions are correct to *first order* in the electron density. Thus a first-order CI wavefunction for the open-shell atoms B, C, N, O, and F would be of quality comparable to the single configuration Hartree–Fock wavefunction for a true closed-shell atom such as neon. Note in this context that Be would not be considered a 'true' closed-shell system because its unoccupied 2p orbital would be extensively utilized in the first-order wavefunction. The formal definition of this new type of wavefunction included all configurations in which at most one electron is assigned to an orbital beyond the valence shell. Such a definition includes higher than double excitations, transcending the internal and semi-internal spaces introduced by Silverstone and Sinanoglu. For molecules[a] the first-order wavefunction is distinguished much more radically from the SS picture. This simple wavefunction was intended to variationally incorporate the structure-sensitive part of the correlation energy into a compact CI form.

[a] H. F. Schaefer, Ph.D. thesis, Stanford University, April, 1969.

69

T. H. DUNNING, W. J. HUNT, & W. A. GODDARD

The theoretical description of the $(\pi\pi^*)$ excited states of ethylene

Chem. Phys. Lett. **4**, 147 (1969)

The $\pi \rightarrow \pi^*$ excited singlet states of conjugated organic molecules have represented a difficult challenge to theoretical chemists for some time. In simple valence electron theories such as Pariser–Parr–Pople, the π^* orbital is expected to be an antibonding combination of atomic $2p\pi$ orbitals, and to have a spatial extent comparable to that of the analogous π orbital. Here Dunning, Hunt, and Goddard showed that the restricted Hartree–Fock method predicts the $\pi \rightarrow \pi^*$ excited singlet state of ethylene to have a π^* orbital which is essentially Rydberg-like, i.e., quite diffuse or spatially extended. Specifically, these authors found the expectation value $\langle x^2 \rangle$ to be 51 atomic units, compared to 12 atomic units for the lowest triplet state (also $\pi \rightarrow \pi^*$ in qualitative designation). Without the inclusion of diffuse basis functions, the $\pi \rightarrow \pi^*$ singlet state is predicted to lie at 9.1 eV, far above the observed 7.6 eV.

More recent theoretical studies[a,b,c] have shown that the explicit introduction of correlation effects results in a more nearly conventional description (i.e., more nearly valence-like) of the $\pi \rightarrow \pi^*$ excited singlet state. However a very delicate balance is involved here and rather flexible variational wavefunctions are required for a definitive result. The most reliable estimate at present is that $\langle x^2 \rangle$ for the $\pi \rightarrow \pi^*$ open-shell singlet state of ethylene is ~20 atomic units. This implies that the Hartree-Fock description of such electronic states is of limited value.

[a] L. E. McMurchie and E. R. Davidson, Configuration interaction calculations on the planar $^1(\pi, \pi^*)$ state of ethylene, *J. Chem. Phys.* **66**, 2959 (1977).
[b] B. R. Brooks and H. F. Schaefer, N(^1A_g), T($^3B_{1u}$), and V($^1B_{1u}$) states of vertical ethylene, *J. Chem. Phys.* **68**, 4839 (1978).
[c] R. J. Buenker, S. -K. Shih, and S. D. Peyerimhoff, An MRD-CI study of the vertical $^1(\pi, \pi^*)$ V-N transition of ethylene using an AO basis with optimized Rydberg $nd\pi$ species and two separate carbon d polarization functions, *Chem. Phys.* **36**, 97 (1979).

70

R. K. NESBET, T. L. BARR, & E. R. DAVIDSON

Correlation energy of the neon atom

Chem. Phys. Lett. **4**, 203 (1969)

Nesbet's earlier-discussed studies of the independent electron pair approach (IEPA) yielded between 98.5 and 100.5 per cent of the correlation energy for the atoms beryllium through neon. Since rather large basis sets were used, this work provided what appeared to be compelling evidence for the validity of the assumptions underlying the IEPA. However, within two years, two groups independently reported a significant discrepancy between energies obtained via the IEPA and variational methods.[a,b] Specifically it was shown that the sum of independently evaluated pair correlation energies is well in excess of the CISD (configuration interaction including all single and double excitations) energy obtained with the same basis set. With this background, Nesbet, Barr, and Davidson revisited the neon atom with a much larger basis set, specifically including functions of higher orbital angular momentum (up to $l = 6$) than previously. With their (8s 6p 5d 4f 3g 2h 1i) Slater basis set, the sum of the pair correlation energies was determined to be 105.2 per cent of the experimental value. Although this result may not today appear terribly shocking, one must realize that the IEPA at the time was expected by many to be reliable to ±1 per cent of the correlation energy.

[a] J. W. Viers, F. E. Harris, and H. F. Schaefer, Pair correlations and the electronic structure of neon, *Phys. Rev. A* **1**, 24 (1970).
[b] T. L. Barr and E. R. Davidson, Nature of the configuration-interaction method of *ab initio* calculations. I. Neon ground state, *Phys. Rev. A* **1**, 644 (1970). Note that although the short communication by Nesbet, Barr, and Davidson appeared in the literature before papers a. and b., the latter papers were in fact widely discussed prior to Nesbet's work.

71

L. C. SNYDER & H. BASCH

Heats of reaction from self-consistent field energies of closed-shell molecules

J. Amer. Chem. Soc. **91**, 2189 (1969)

By this time it was well established that the Hartree–Fock approximation provided poor values for the dissociation energies of diatomic molecules. Many distinguished quantum mechanicians were of the opinion that SCF theory was going to be virtually useless in the prediction of thermochemical data. In this framework, Snyder and Basch came to the perhaps surprising conclusion that certain kinds of thermodynamic properties were quite reasonably treated within the Hartree–Fock approximation. Specifically, they used Whitten's double zeta

(DZ) contracted gaussian basis sets to systematically describe a family of 26 closed-shell molecules expected to be well-described by a single configuration wavefunction. For reactions involving only these closed-shell molecules, the predicted heats of reaction are in good qualitative agreement with experiment, the differences being typically less than 10 kcal/mole. More recently Pople and his co-workers have narrowed the Snyder–Basch concept to *isodesmic* reactions and achieved considerable success in its application.[a] An isodesmic reaction is one for which the number of each type of bond (e.g., C=C) is identical for reactants and products.

[a] W. J. Hehre, R. Ditchfield, L. Radom, and J. A. Pople, Molecular orbital theory of the electronic structure of organic compounds. V. Molecular theory of bond separation, *J. Amer. Chem. Soc.* **92**, 4796 (1970).

72

R. C. LADNER & W. A. GODDARD

Improved quantum theory of many-electron systems. V. The spin-coupling optimized GI method

J. Chem. Phys. **51**, 1073 (1969).

We have earlier noted Löwdin's 1955 proposal of the 'extended Hartree–Fock' method, wherein orbital optimization was to be carried out *after* the application of a spin projection operator to a Slater determinant with n spatially distinct orbitals, where n is the number of electrons. Goddard took the kernel of Löwdin's idea and carried it through from mathematical development to practical implementation.[a,b] In Goddard's formalism, the different spin functions associated with n unpaired orbitals are represented by Young tableaux and are designated $G1$, $G2$, $G3$, . . . GF, the last function GF being identical to the extended Hartree–Fock wavefunction. Goddard noted that an even better wavefunction is obtained by taking a linear combination of the functions $G1$, $G2$, $G3$, . . . GF and minimizing the energy *both* with respect to the molecular orbitals and the configuration mixing coefficients. The latter method was designated the spin-coupling optimized GI (or SOGI) method and the first applications to molecular systems (LiH, H_3, H_4) were presented here. The $G1$ or valence bond-like coupling was found to dominate the overall wavefunction.

[a] W. A. Goddard, Improved quantum theory of many-electron systems. I. Construction of eigenfunctions of $\hat{S}^2$ which satisfy Pauli's principle, *Phys. Rev.* **157**, 73 (1967).

[b] W. A. Goddard, Improved quantum theory of many-electron systems. II. The basic method, *Phys. Rev.* **157**, 81 (1967).

73

P. PULAY

Ab initio calculation of force constants and equilibrium geometries in polyatomic molecules I. Theory

Mol. Phys. **17**, 197 (1969)

If one may distinguish among landmark papers, then Pulay's early work[a] on the analytic gradient method must be said to comprise a 'super landmark'. In the early sections of the paper, Pulay emphasized the unreliability of the Hellman-Feynman method for evaluating forces on nuclei. Although the latter method is exact for true Hartree-Fock wavefunctions and trivial to apply, it is found to be of little value for the finite basis sets typically used in molecular calculations. Pulay's proposal was to determine the forces (energy first derivatives) analytically, requiring the explicit evaluation of the first derivatives of the one- and two-electron integrals. However, this is quite straightforward for the gaussian lobe function basis sets used in the early work of Pulay. Force constants were then evaluated as finite differences of analytic gradients. During the 1970s Pulay (often in collaboration with Wilfried Meyer) reported a stream of impressive chemical applications[b] of what he termed 'the force method'.

[a] The next two papers in the series are: P. Pulay, *Ab initio* calculations of force constants and equilibrium geometries in polyatomic molecules. II. Force

constants of water, *Mol. Phys.* **18**, 473 (1970); III. Second row hydrides, *Mol. Phys.* **21**, 329 (1971).

[b] See, for example, P. Pulay and W. Meyer, Comparison of *ab initio* force constants of ethane, ethylene, and acetylene, *Mol. Phys.* **27**, 473 (1974).

74

U. KALDOR & F. E. HARRIS

Spin-optimized self-consistent field wave functions

Phys. Rev. **183**, 1 (1969)

The first fully self-consistent *ab initio* results for the extended Hartree–Fock model of Löwdin were obtained by Goddard for several simple systems[a] and followed within a year by Kaldor's results[b] for the lithium atom. These two groups were also involved in a competition to develop what Goddard had called the SOGI method. The latter method was designated the spin-optimized SCF approach by Kaldor and Harris, who used an extended Brillouin's theorem to solve the resulting equations. The valence bond-like spin function labeled *G*1 by Goddard was called the maximally paired (MP) SCF wavefunction by Kaldor and Harris. In the present writer's mind, the important point is not the rivalry between nomenclatures but the fact that both groups made seminal contributions to the development of theoretical chemistry. The specific systems considered here by Kaldor and Harris include the Li and Be atom ground states, both of which have two linearly independent spin functions. One of the more difficult problems solved later by the SOSCF method was the nitrogen atom ground state.[c] For that seven electron system there are six independent ^4S spin eigenfunctions and Kaldor found the SOSCF method to account for 11 per cent of the correlation energy.

[a] Note previously cited papers and W. A. Goddard, Magnetic hyperfine structure of lithium, *Phys. Rev.* **157**, 93 (1967).

[b] U. Kaldor, Calculation of extended Hartree–Fock wave functions, *J. Chem. Phys.* **48**, 835 (1968).

[c] U. Kaldor, Spin-optimized self-consistent-field function. II. Hyperfine structure of atomic nitrogen, *Phys. Rev. A* **1**, 1586 (1970).

75

C. F. BENDER & E. R. DAVIDSON

Studies in configuration interaction: The first-row diatomic hydrides

Phys. Rev. **183**, 21 (1969)

I must confess to a certain level of astonishment when I came upon this manuscript in preprint form in early 1969. Although the Nesbet method for the extraction of the lowest eigenvalue (and eigenvector) from a large, real, symmetric matrix had been introduced four years earlier, most electronic structure theorists considered 400 configurations to be an exceptionally 'large' CI. Bender and Davidson shattered this illusion by reporting much more complete wavefunctions for the entire series of diatomic hydrides LiH through HF. For example, for the NH molecule a Slater basis set of size N(14σ 13π 6δ 1ϕ) H(3σ 3π 3δ 1ϕ) was used in conjunction with a CI of 3379 configurations to yield 74 per cent of the experimental correlation energy. The iterative natural orbital procedure (no more than three iterations) was used to guarantee that the molecular orbital basis set was close to optimum. Dipole moments were also predicted at the NO-CI level and excellent agreement with experiment (in parentheses) was found for μ(LiH) = 5.85(5.82), μ(CH) = 1.43(1.4), μ(OH) = 1.63(1.66) and μ(FH) = 1.82(1.82) debyes, the four cases for which experimental values were available. The success of the Bender-Davidson paper clearly established large CI as the method of choice for the immediate future.

76

R. P. HOSTENY, R. R. GILMAN, T. H. DUNNING, A. PIPANO, & I. SHAVITT

Comparison of Slater and contracted Gaussian basis sets in SCF and CI calculations on H_2O

Chem. Phys. Lett. **7**, 325 (1970)

In spite of their widespread adoption by the end of 1970, contracted gaussian basis sets were still looked upon with considerable suspicion by some of the afficionados of molecular quantum mechanics. This paper by Hosteny *et al.* broke the spell by offering a direct comparison for the water molecule between the respectable Slater functions and the upstart contracted gaussians. For Slater functions, the standard STO double zeta basis set of Clementi was used, while for gaussians the new Huzinaga–Dunning double zeta contraction was adopted. Thus both sets involved precisely the same number of final functions. Remarkably, the gaussian basis set yielded a lower total energy at the SCF level and with four different types of CI. The energy differences (0.003–0.007 hartree) were not large, so the general conclusion was that the two types of basis sets are essentially comparable. However, the computation time required for the Slater basis was 569 seconds (CDC 6400) while that needed for the gaussian set was only 49 seconds on the same machine. Gaussian functions were here to stay.

77

A. K. Q. SIU & E. R. DAVIDSON

A study of the ground state wave function of carbon monoxide

Int. J. Quantum Chem. **4**, 223 (1970)

If the Nesbet–Barr–Davidson paper dealt a severe blow to the independent electron pair approach (IEPA), then it is probably fair to say that

Siu and Davidson delivered the knockout punch. For the CO molecule they used a large basis set (18σ 15π 6δ 4ϕ) and reported an SCF energy 0.008 atomic units above McLean and Yoshimine's near Hartree-Fock result. After four natural orbital iterations, the authors settled on a 2484 configuration variational wavefunction, accounting for 70 per cent of the correlation energy. Pair correlation energies were also reported, but the sum of these is a much greater fraction (93 per cent) of the experimental correlation energy. The problem, clearly articulated by Siu and Davidson, is that because IEPA 'ignores matrix elements between configurations belonging to different pairs, it tends to overestimate the correlation energy when several orbitals occupy the same region of space'. The triple bond in CO is obviously an example of such a situation. Without contesting the qualitative insights afforded by the IEPA, this paper in essence put an end to it as a quantitative method for the prediction of total energies.

78

C. F. BENDER & H. F. SCHAEFER

New theoretical evidence for the nonlinearity of the triplet ground state of methylene

J. Amer. Chem. Soc. **92**, 4984 (1970)

By this time Boys's 1960 prediction of a 129° bond angle for triplet methylene had been virtually fogotten. Herzberg's experimental determination of linearity was buttressed by the highly influential semi-empirical theoretical study of Longuet-Higgins,[a] then Professor of Theoretical Chemistry at Cambridge. Moreover, while several theoretical predictions of a bent triplet methylene were made in the 1960s, these frequently sought to rationalize why they did not achieve the 'correct' result, namely a linear structure. The paper at hand has been described by Gaspar and Hammond[b] as 'by far the most elaborate calculation carried out to that date on methylene, or indeed almost any molecule'. An equilibrium bond angle θ(HCH) = 135.1° was predicted using these sophisticated techniques. Moreover, the authors were convinced that Herzberg's interpretation of his spectroscopic data was incomplete, and pointed to an alternative spectroscopic assignment suggested but discarded by Herzberg. Within months this apparent discrepancy between theory and experiment was resolved in favour of the former by the electron spin resonance studies of Bernheim and Skell at Penn State and Wasserman at Bell Laboratories.

[a] P. C. H. Jordan and H. C. Longuet-Higgins, The lower electronic levels of the radicals CH, CH_2, CH_3, NH, NH_2, BH, BH_2, and BH_3, *Mol. Phys.* **5**, 121 (1962).

[b] P. P. Gaspar and G. S. Hammond, Spin states in carbene chemistry, pages 207–362 of Volume II, *Carbenes,* editors R. A. Moss and M. Jones (Wiley, New York, 1975).

79

D. M. SILVER, K. RUEDENBERG, & E. L. MEHLER

Electron correlation and separated pair approximation in diatomic molecules. III. Imidogen

J. Chem. Phys. **52**, 1206 (1970)

Early work on the separated pair approximation inevitably leads to the classic 1953 paper of Hurley, Lennard-Jones, and Pople.[a] Thence comes the idea that the wavefunction might be approximated by an antisymmetrized product of strongly orthogonal geminals (APSG). The term 'strongly orthogonal' means that the two-electron functions (geminals) ω_R and ω_S satisfy the relationship

$$\int \omega_R(1,2)\, \omega_S(1,3)\, d\tau_1 = 0 \text{ for } R \neq S.$$

Ruedenberg's work showed that it was indeed possible to determine accurate separated pair wavefunctions by strictly *ab initio* methods. The separated pairs thus obtained are indeed found to describe the sort of localized electron pairs anticipated by G. N. Lewis early in this century. Moreover the APSG wavefunction is size-consistent, i.e., the total correlation energy is proportional to n. The largest system studied by the Ruedenberg group was the NH molecule, many properties of which were predicted. The separated pair dissociation energy is found for example to be 2.65 eV, significantly less than experiment, ~3.4 eV. The authors concluded that the APSG 'is too restrictive to yield a complete description of all the various electron correlations in this molecular system'. The reason for this of course is the neglect of the inter-pair correlation effects and to a lesser degree the severity of the strong orthogonality constraint.

[a] A. C. Hurley, J. E. Lennard-Jones, and J. A. Pople, The molecular orbital theory of chemical valency. XVI. A theory of paired electrons in polyatomic molecules, *Proc. Roy. Soc. (London)* **A220**, 446 (1953).

80

M. D. NEWTON, W. A. LATHAN, W. J. HEHRE, & J. A. POPLE

Self-consistent molecular orbital methods. V. *Ab initio* calculation of equilibrium geometries and quadratic force constants

J. Chem. Phys. **52**, 4064 (1970)

The first systematic study of the geometrical structures of polyatomic molecules is presented here. In addition this paper was one of the earliest based on the now-famous GAUSSIAN 70 program developed at Carnegie-Mellon University. The method employed (minimum basis set SCF theory with each Slater function replaced by a linear combination of three primitive gaussians (STO-3G)) was one of the simplest imaginable, and it was not at all obvious to most workers at the time that such an approach would yield meaningful geometries. However, most of the structures predicted (30 species with more than 75 unique bond distances and angles) were in good qualitative agreement with experiment. The widely cited exceptions were the FOOF and FNNF molecules, for which discrepancies as large as 0.2 Å (in bond distances) relative to experiment were found. A general conclusion of this and later studies is that structures involving heteroatoms (N, O, F) are unreliable at this level of theory (MBS SCF), while hydrocarbon structures are remarkably close to experiment. It should perhaps be added that the FOOF molecular structure continues in mediocre agreement with experiment even as one approaches the Hartree–Fock limit of basis set completeness.[a]

[a] R. R. Lucchese, H. F. Schaefer, W. R. Rodwell, and L. Radom, Fluorine peroxide (FOOF): a problem molecule for theoretical structural predictions, *J. Chem. Phys.* 68, 2507 (1978).

81

P. J. BERTONCINI, G. DAS, & A. C. WAHL

Theoretical study of the $^1\Sigma^+$, $^3\Sigma^+$, $^3\Pi$, $^1\Pi$ states of NaLi and the $^2\Sigma^+$ state of $NaLi^+$

J. Chem. Phys. 52, 5112 (1970)

Wahl's method of 'optimized valence configurations' involved a small (typically less than 20 configurations) MCSCF wavefunction designed[a] to treat 'only that part of the system whose correlation energy changes considerably during molecular formation'. Certainly one of the most important early applications of the method was to the NaLi molecule. This was a purely 'blind' test of the theory since NaLi was in 1970 the only light diatomic alkali molecule which had not been observed in the laboratory. Predictions were made for the $^1\Sigma^+$, $^3\Sigma^+$, $^3\Pi$, and $^1\Pi$ potential curves of NaLi and for the $^2\Sigma^+$ ground state of the positive ion. Most elaborate were the ground state MCSCF calculations, where the significance of six important configurations was established. Qualitative features of the NaLi chemical bond were beautifully illustrated via contour diagrams. Perhaps the most important ground state predictions were r_e(Na–Li) = 2.936 Å, D_e = 0.85 eV and ω_e = 248 cm^{-1}. Subsequently NaLi was indeed observed spectroscopically,[b] and it was found that r_e(Na–Li) = 2.826 Å and ω_e = 257 cm^{-1}.

[a] A. C. Wahl and G. Das, The method of optimized valence configurations: a reasonable application of the MCSCF technique to the quantitative description of chemical bonding, *Adv. Quantum Chem.* **5**, 261 (1970).
[b] M. M. Hessel, Experimental observation of the NaLi molecule, *Phys. Rev. Lett.* **26**, 215 (1971).

82

H. F. SCHAEFER

New approach to electronic structure calculations for diatomic molecules: application to F_2 and Cl_2

J. Chem. Phys. **52**, 6241 (1970)

Although the valence bond method played an important role in early qualitative theories of chemical bonding, the presence of non-orthogonal atomic orbitals (i.e., not orthogonal to atomic crbitals on other nuclear centres) made this approach difficult to implement in *ab initio* formulations of the theory. However, the present paper presented a strictly *numerical* method for the evaluation of two-centre integrals and thus allowed the direct use of Hartree–Fock atomic orbitals of arbitrary precision. Although these Hartree–Fock orbitals were subsequently orthogonalized, the use of a full CI (within the space of the occupied atomic orbitals) guaranteed results entirely equivalent to a complete *ab initio* valence bond treatment. For F_2 and Cl_2 the dissociation energies predicted in this manner were 0.32 and 0.71 eV, respectively, compared to experiment, ~1.6 eV and 2.48 eV. It was clear that the true Hartree–Fock atomic orbitals are too inflexible to be used profitably in valence bond studies. The methods and programs developed here also proved useful in the study of a number of systems for which one electron must be described as a continuum orbital.[a,b]

[a] P. K. Pearson and H. Lefebvre-Brion, Calculations of the widths of some Feschbach resonances in CO^- and NO^-, *Phys. Rev. A.* **13**, 2106 (1976).
[b] K. C. Kulander and J. S. Dahler, A theory of transfer ionization: application to $He^+ + Mg \rightarrow He + Mg^{+2} + e^-$, *J. Phys. B* **8**, 460 (1975).

83

A. BUNGE

Electronic wave functions for atoms. III. Partition of degenerate spaces and ground state of C

J. Chem. Phys. **53**, 20 (1970)

An important new concept for the treatment of electron correlation in open-shell systems arose from this study. For such atomic systems, a given orbital occupancy (or electron configuration) can give rise to several distinct *L–S* eigenfunctions. For example, for the carbon atom ground state ($1s^2 2s^2 2p^2$) the double excitation 1s2s → 3s4s yields the orbital occupancy

$$1s\ 2s\ 3s\ 4s\ 2p^2,$$

for which there are *six* different ^{3}P *L–S* eigenfunctions. However, Bunge showed that it is possible to partition this six-fold degenerate spin space in such a way that only two configurations, designated via intermediate coupling as

$$(1s\ 2s\ 3s\ 4s)\ {}^1S\ 2p^2$$

are used variationally, while the other four are found to have identically zero matrix elements with the Hartree–Fock reference configuration. Furthermore, Bunge showed that configurations of the type neglected were of little importance to the wavefunction or energy. The validity of this approach was quickly verified for molecules[a] and more recently has been extended somewhat.[b,c]

[a] C. F. Bender and H. F. Schaefer, Electronic splitting between the $^2B_1$ and $^2A_1$ states of the NH_2 radical, *J. Chem. Phys.* **55**, 4798 (1971).

[b] A. D. McLean and B. Liu, Classification of configurations and the determination of interacting and noninteracting spaces in configuration interaction, *J. Chem. Phys.* **58**, 1066 (1973).

[c] K. Iberle and E. R. Davidson, Integral dependent spin couplings in CI calculations, *J. Chem. Phys.* **76**, 5385 (1982).

84

J. M. SCHULMAN & D. N. KAUFMAN

Application of many-body perturbation theory to the hydrogen molecule

J. Chem. Phys. **53**, 477 (1970)

Hugh Kelly[a] and later the group of T. P. Das[b] demonstrated clearly during the 1960s that many-body perturbation theory (MBPT) was a powerful approach to the electronic structure of atoms. However for molecules, where the use of anything approaching a complete set of numerical functions was impractical, the situation was uncertain. Clearly what was required was an adaptation of MBPT to the analytic Slater and gaussian basis sets used so successfully in variational studies. First steps in that direction were taken in the present paper by Schulman and Kaufman. Although the system chosen, the hydrogen molecule, was very simple, a rather thorough study, including predictions of total energy, electric dipole polarizabilities, and the electron-coupled nuclear spin-spin Fermi interaction was completed. An uncontracted gaussian basis set of size H(10s 5p 1d) was chosen and tested for practical completeness in several ways. Satisfactory results were obtained, completing the initial stage of the use of analytic basis sets in MBPT.

[a] H. P. Kelly, Applications of many-body diagram techniques in atomic physics, *Adv. Chem. Phys.* **14**, 129 (1969).
[b] N. C. Dutta, C. Matsubara, R. T. Pu, and T. P. Das, Many-body approach to hyperfine interaction in atomic nitrogen, *Phys. Rev.* **177**, 33 (1969).

85

T. H. DUNNING

Gaussian basis functions for use in molecular calculations. I. Contraction of (9s 5p) atomic basis sets for the first-row atoms

J. Chem. Phys. **53**, 2823 (1970)

It is reasonable to contend that prior to the present paper all contractions of sizeable gaussian basis sets resulted in considerable loss in the calculated SCF total energy relative to the use of the full primitive basis. Without taking away from the pioneering efforts of earlier workers, it seems fair to say that Dunning's paper set a new standard for the systematic and effective contraction of gaussian basis sets. Among the general principles set down by Dunning were: (a) functions of greatest importance in regions between the nuclei should be left uncontracted; (b) functions spanning two different orbital spaces (e.g., 1s and 2s of a particular atom) should also be allowed to vary freely. Using these guidelines Dunning was able to contract Huzinaga's (9s 5p) basis for the atoms B through F to (4s 2p) with relatively little increase in the SCF energy. For example, for the fluorine atom the uncontracted and contracted energies were −99.3956 and −99.3933 hartrees, the difference being only 0.0023 hartree. These contracted sets and the later developed (10s 6p/5s 3p) bases[a] have been widely used in electronic structure theory during the past decade.

[a] T. H. Dunning, Gaussian basis functions for use in molecular calculations. III. Contraction of (10s 6p) atomic basis sets for the first row atoms, *J. Chem. Phys.* **55**, 716 (1971).

86

H. F. SCHAEFER,
D. R. McLAUGHLIN,
F. E. HARRIS, & B. J. ALDER

Calculation of the attractive He pair potential

Phys. Rev. Lett. **25**, 988 (1970)

The following two papers reported simultaneously the first qualitatively successful *ab initio* predictions of a van der Waals interaction. The term 'successful' is used in the sense of achieving the 'right answer for the right reason'. In quantum mechanics, of course, it is always possible to accidentally make a prediction in agreement with experiment, but using a woefully inadequate theoretical method. The first of the two papers begins with the realization that a simple transformation of the He_2 Hartree-Fock wavefunction to localized orbitals yields a wavefunction $A(4)\ 1\sigma_A^2\ 1\sigma_B^2$, in which $1\sigma_A$ and $1\sigma_B$ are slightly distorted atomic 1s orbitals on helium atoms A and B. In this picture it is seen that the inter-atomic correlation effects are represented by excitations of the type $1\sigma_A\ 1\sigma_B \rightarrow xy$. To some low order, then, the potential energy curve is well approximated by the sum of the Hartree-Fock energy plus the interatomic correlation contributions. Equally important, this paper sheds light on the origin of the van der Waals attraction by dissecting the attraction into contributions from s, p, d, and f atomic orbitals and into σ, π, δ, and ϕ molecular orbital terms.

87

P. J. BERTONCINI & A. C. WAHL

Ab initio calculation of the helium–helium $^1\Sigma_g^+$ potential at intermediate and large separations

Phys. Rev. Lett. **25**, 991 (1970)

The second landmark paper on the He-He attraction made effective use of MCSCF procedures. The localized Hartree-Fock wavefunction $1\sigma_A^2\ 1\sigma_B^2$ was retained throughout, but eight additional configurations of the type $1\sigma_A\ 1\sigma_B \rightarrow xy$ were added and optimum molecular orbitals determined via the MCSCF method. The predicted dissociation energy D_e(He-He) was 11.4 K, while it was estimated that the extension of the basis set would probably converge to a well depth of close to 12 K. Not surprisingly, the previous paper, using a larger basis set, determined a binding energy of 12.0 K. Thus the two methods, which should be essentially equivalent, yielded very similar results. The experimental well depth is now known[a] to be ~10.6 K, so the neglect of intra-atomic correlation effects is seen to overestimate the binding energy by ~10 per cent. This method, although based on a well-defined theoretical picture, appears to be of less value for larger van der Waals systems. Even for such closely related systems Ne-Ne or He-H_2, the assumption of constant intra-atomic correlation energy as a function of geometry is significantly less valid than for He-He.[b]

[a] A. L. J. Burgmans, J. M. Farrar, and Y. T. Lee, Attractive well of He-He from ^{3}He-^{4}He differential elastic scattering measurements, *J. Chem. Phys.* **64**, 1345 (1976).

[b] W. Meyer, P. C. Hariharan, and W. Kutzelnigg, Refined *ab initio* calculation of the potential energy surface of the He—H_2 interaction with specific emphasis to the region of the van der Waals minimum, *J. Chem. Phys.* **73**, 1880 (1980).

88

T. H. DUNNING & N. W. WINTER

Hartree–Fock calculation of the barrier to internal rotation in hydrogen peroxide

Chem. Phys. Lett. **11**, 194 (1971)

As early as Pitzer and Lipscomb's classic 1963 paper on the internal rotation barrier in ethane, evidence began to accumulate that conformational energy differences were reliably predicted within the Hartree-Fock approximation. However, the HOOH molecule remained a serious problem until 1971. Hydrogen peroxide has a gauche (point group C_2) equilibrium geometry, with maxima (barriers) occurring for both the planar *cis* (~7.6 kcal) and planar *trans* (1.1 kcal) stationary points. Previous theoretical studies had difficulty reproducing either of these experimental barriers. Dunning and Winter concluded that both polarization functions (oxygen d functions plus hydrogen p functions) and careful geometrical optimization of all three stationary points were required for a meaningful study of the conformational energy surface. In this manner, barriers of 8.4 kcal (*cis*) and 1.1 kcal (*trans*) were predicted at the SCF level of theory. In addition, Dunning and Winter had the audacity to suggest that their predicted *cis* barrier might actually be more reliable than the experimentally derived result. Thus for the time being Dunning and Winter put to rest the last apparent discrepancy between SCF theory and experiment for a conformational problem. However, very recently it has become likely[a] that the barrier to heavy atom linearity in propadienone (CH_2=C=C=O) is not treated in a qualitatively correct manner as one approaches the Hartree-Fock limit.

[a] L. Farnell and L. Radom, The structure of propadienone (CH_2=C=C=O), *Chem. Phys. Lett.* **91**, 373 (1982).

89

F. GREIN & T. C. CHANG

Multiconfiguration wavefunctions obtained by application of the generalized Brillouin theorem

Chem. Phys. Lett. **12**, 44 (1971)

In an important paper appearing in 1968, Levy and Berthier[a] formulated a generalized Brillouin condition for multiconfiguration wavefunctions. Ruedenberg and co-workers[b] (who also make use of these ideas) have recently designated this condition the Brillouin–Levy–Berthier (BLB) theorem. The essence of the BLB condition is that for variationally optimum multiconfiguration (MC) wavefunctions, one finds

$$\langle \psi_{\mathrm{MC}} | \mathrm{H} | \psi_{\mathrm{MC}}(i \rightarrow j) \rangle = 0,$$

where $\psi_{\mathrm{MC}}(i \rightarrow j)$ implies that the single excitation $i \rightarrow j$ has been properly implemented for *each* configuration included in the MC wavefunction under consideration. The first *ab initio* application of the BLB ideas was probably that of O'Neil and Bender[c] in their two-configuration SCF treatment of singlet methylene. However, the first systematic and theoretically precise implementation was reported in the present paper by Grein, who investigated the He, Li, and Be atom ground states. In addition to Grein and Rudenberg, Yarkony[d] has developed and applied a formalism based on the BLB concept in recent years.

[a] B. Levy and G. Berthier, Generalized Brillouin theorem for multiconfigurational SCF theories, *Int. J. Quantum. Chem.* **2**, 307 (1968).

[b] K. Ruedenberg, L. M. Cheung, and S. T. Elbert, MCSCF optimization through combined use of natural orbitals and the Brillouin–Levy–Berthier theorem, *Int. J. Quantum Chem.* **16**, 1069 (1979).

[c] S. V. O'Neil, H. F. Schaefer, and C. F. Bender, C_{2v} potential energy surfaces for seven low-lying states of CH_2, *J. Chem. Phys.* **55**, 162 (1971).

[d] D. R. Yarkony, Symmetry-adapted-multiconfiguration SCF wavefunctions via symmetry-restricted annihilation of single excitations. I., *J. Chem. Phys.* **66**, 2045 (1977).

90

W. MEYER

Ionization energies of water from PNO–CI calculations

Int. J. Quantum Chem. Symp. **5**, 341 (1971)

As noted earlier, an obvious goal of theoretical chemistry during the 1960s was to develop a method for many-electron systems that would embody the remarkable two-electron convergence properties characteristic of natural orbitals. Thus, for instance, in their formulation of the IEPA, Kutzelnigg and co-workers required that each independent pair correlation function be expressed in terms of its own (approximate) pair natural orbitals (PNOs). For variational n-electron studies, of course, these different sets ($n^2/4$ in all) of pair natural orbitals are mutually non-orthogonal and therefore assumed not suitable for CI studies. However, Meyer challenged this assumption, noting that each member of a particular set of PNOs was orthogonal to the other members and to the occupied SCF orbitals. Thus the non-orthogonality problem was not as gruesome as one might have expected,[a] and Meyer was able to formulate the PNO–CI method. His ground state wavefunction for water thus consisted of only ~240 configurations, but these were constructed from ~350 different pair natural orbitals. About 86 per cent of the H_2O correlation energy was obtained in this manner, a variational result which is yet to be improved upon.

[a] See Appendix A in the paper by G. G. Balint-Kurti and M. Karplus, Multistructure valence-bond and atoms-in-molecules calculations for LiF, F_2, and F_2^-, *J. Chem. Phys.* **50**, 478 (1969).

91

E. CLEMENTI, J. MEHL, & W. VON NIESSEN

Study of the electronic structure of molecules. XII. Hydrogen bridges in the guanine–cytosine pair and in the dimeric form of formic acid

J. Chem. Phys. **54**, 508 (1971)

Many theoretical chemists have a conscious or subconscious desire to study the electronic structure of DNA, understanding as we do that this substance is quite central to why human beings are as we are. The first concrete *ab initio* step in this direction was taken in 1971 by Clementi's group. They considered the guanine–cytosine base pair, which is held together by no less than three hydrogen bonds. A gaussian basis of size C, N, O(7s 3p), H(3s) was contracted to minimal size in conjunction with the SCF method. Although the total number of contracted functions (105) seems not excessive by 1983 standards, it was staggering at the time of publication. To understand the interplay between the three hydrogen bonds, 27 different guanine–cytosine geometrical arrangments were considered, requiring eight days of computation on the IBM 360/195, then the fastest (or perhaps second only to the CDC 7600) machine in the world. As a check on the reliability of the quantum mechanical predictions, parallel studies (including higher levels of theory) were reported for the much simpler formic acid dimer, which has two hydrogen bonds. Both of these problems could be profitably reinvestigated with modern analytic derivative methods, but the work of Clementi represented a genuine *tour de force* for 1971.

92

H. F. SCHAEFER

Ab initio potential curve for the $X^3\Sigma_g^-$ state of O_2

J. Chem. Phys. **54**, 2207 (1971)

This paper represents the first direct application to molecules of the ideas introduced by Silverstone and Sinanoglu in 1966 and extended by Schaefer and Harris in 1968 in their formulation of the 'first-order' wavefunction. To illustrate the difference between these two methods, one may note the example of the ethylene radical cation $C_2H_4^+$ given in the second 1966 SS paper.[a] SS state that for $C_2H_4^+$ there are no significant internal or semi-internal correlation effects, but that the single excitations (spin polarization in their syntax) should be included in their model. Taken as a whole, the enumerated contributions to the correlation energy of $C_2H_4^+$ are essentially miniscule. A very different picture appears in the first-order wavefunction, since it defines the six lowest virtual orbitals of $C_2H_4^+$ to lie within the valence shell, defined as that orbital space pictorially spanned by the carbon 2s, $2p_x$, $2p_y$, $2p_z$ and hydrogen 1s atomic orbitals. The first-order dictum to include all configurations in which not more than one electron occupies an orbital beyond the valence shell thus includes a very large number of important configurations (all ignored by SS) and recovers a large fraction of the valence-shell correlation energy. The present paper applies these ideas successfully to the O_2 molecule, exploiting the iterative natural orbital method to properly shape the valence orbitals not occupied in the SCF wavefunction.

[a] H. J. Silverstone and O. Sinanoglu, Many-electron theory of nonclosed-shell atoms and molecules. II. Variational theory, *J. Chem. Phys.* **44**, 3608 (1966).

93

S. ROTHENBERG &
H. F. SCHAEFER

Methane as a numerical experiment for polarization basis function selection

J. Chem. Phys. **54**, 2764 (1971)

After the different nuclei, the next most obvious place in which to locate analytic basis functions is along the line of centres between two nuclei.[a] The first systematic study of such 'bond functions' is provided for the methane molecule in the present paper. In particular it was demonstrated that bond functions play a role analogous to polarization basis functions (for CH_4 the first polarization functions are carbon d and hydrogen p functions). Experimentation was carried out with respect to the positions and scale factors associated with spherical gaussian (i.e., 1s) bond functions. It was found that eight optimized bond functions (two located along each bond) are more effective in lowering the SCF energy than is an optimized set of carbon d functions. Later studies showed that p functions located along bonds could also be effective as polarization functions.[b] A philosophical objection to bond functions arises in reactive systems, where the number of 'bonds' can become ill-defined, and accordingly the placement of bond functions takes on a measure of arbitrariness.

[a] For an early theoretical study including bond functions, see R. Ahlrichs and W. Kutzelnigg, *Ab initio* calculations of small hydrides including electron correlation. II. Preliminary results for the CH_4 ground state, *Chem. Phys. Lett.* 1, 651 (1968).

[b] T. Vladimiroff, Comparison of the use of 3d polarization functions and bond functions in gaussian Hartree–Fock calculations, *J. Phys. Chem.* 77, 1983 (1973).

94

J. B. ROSE & V. McKOY

Applicability of SCF theory to some open-shell states of CO, N_2 and O_2

J. Chem. Phys. **55**, 5435 (1971)

When it is possible to express the energy of a single configuration wavefunction as a linear combination of integrals

$$h_i = \langle \phi_i | \mathrm{h} | \phi_i \rangle$$

$$J_{ij} = \langle \phi_i(1)\, \phi_j(2) | 1/r_{12} | \phi_i(1)\, \phi_j(2) \rangle$$

$$K_{ij} = \langle \phi_i(1)\, \phi_j(2) | 1/r_{12} | \phi_j(1)\, \phi_i(2) \rangle$$

one can always (barring practical difficulties) solve the resulting Hartree-Fock equations using now standard techniques.[a] However, for molecules having degenerate point group symmetry, many of the low-lying excited states may not fall in this category. For example, the configuration $\ldots 3\sigma_g^2\ 1\pi_u^3\ 1\pi_g$ is low-lying for the N_2 molecule and gives rise to the four Σ states $^1\Sigma_u^+$, $^3\Sigma_u^+$, $^1\Sigma_u^-$, and $^3\Sigma_u^-$. The conventional Hartree-Fock energy expressions for all four of these N_2 states involve more complicated integrals (than J_{ij} or K_{ij}), i.e., integrals involving three or four spatially distinct molecular orbitals. What Rose and McKoy showed was that by using real molecular orbitals π_x and π_y instead of the conventional complex π_+ and π_-, it was possible to transform these energy expressions to involve only the two-index integrals J_{ij} and K_{ij}. Thus the scope of applicability of restricted Hartree–Fock theory was considerably enlarged.

[a] W. J. Hunt, T. H. Dunning, and W. A. Goddard, The orthogonality constrained basis set expansion method for treating off-diagonal Lagrange multipliers in calculations of electronic wavefunctions, *Chem. Phys. Lett.* **3**, 606 (1969).

95

B. ROOS

A new method for large-scale CI calculations

Chem. Phys. Lett. **15**, 153 (1972)

The introduction of the 'direct CI' method by Roos created a revolution in the electronic structure theorist's thinking about how to design correlated wavefunctions. Previously all workers had explicitly constructed all elements H_{ij} of the hamiltonian matrix and then employed some standard (typically Nesbet's) eigenvalue method to determine the energy and wavefunction. However one sees (even if only 1 per cent of the H_{ij} are non-zero) that for a wavefunction of one million configurations, there will be five billion non-vanishing hamiltonian matrix elements. Since no existing computer can comfortably store this amount of information (even on peripheral devices), it is clear that the explicit formation of the hamiltonian matrix cannot be considered practical for the largest CI wavefunctions. For closed-shell CISD (all single and double excitations) Roos showed that it was possible to work directly from the one- and two-electron integrals to the eigenvalue and eigenvector without ever dealing *per se* with the hamiltonian matrix elements. Proof of the effectiveness of the method was given by a CI of 4336 configurations (the largest ever reported at the time) for the ammonia molecule. As we shall see, the direct CI concept is now exploited in one form or another by most state-of-the-art methods for the determination of correlated wavefunctions.

96

J. A. HORSLEY, Y. JEAN, C. MOSER, L. SALEM, R. M. STEVENS, & J. S. WRIGHT

An organic transition state

J. Amer. Chem. Soc. **94**, 279 (1972)

Since the pioneering work of Wigner, Polanyi, and Eyring in the 1930s the transition state concept has played a key role in the understanding of chemical kinetics. Therefore it was not surprising that *ab initio* techniques would eventually be brought to bear on the precise location of transition states for organic reactions. The first such achievement for a sizeable system was that of Salem and his colleagues for the geometrical isomerization of cyclopropane

$$\begin{array}{c} CH_2 \\ / \quad \backslash \\ H_2C - CH_2 \end{array} \longrightarrow \begin{array}{c} CH_2 \\ / \quad \backslash \\ H_2C\cdot \quad \cdot CH_2 \end{array} \longrightarrow \begin{array}{c} CH_2 \\ / \quad \backslash \\ H_2C - CH_2 \end{array}$$

Their breakthrough was all the more striking because the trimethylene transition state is a diradical and not even qualitatively described by a single configuration wavefunction. More specifically, three configurations were required to map out the reaction pathway for this degenerate rearrangement. In the absence of analytic gradient methods,[a] it was a genuinely heroic task to resolve the structure of the trimethylene transition state within the full 21-dimensional energy hypersurface.

[a] The same problem was revisited using gradient techniques by S. Kato and K. Morokuma, Energy gradient in a multi-configurational SCF formalism and its application to geometry optimization of trimethylene diradicals, *Chem. Phys. Lett.* **65**, 19 (1979).

97

L. R. KAHN & W. A. GODDARD

Ab initio effective potentials for use in molecular calculations

J. Chem. Phys. **56**, 2685 (1972)

For decades chemists have known that the observable properties of most molecules are governed largely by their valence electrons. For this reason, *ab initio* theorists have long desired to somehow remove the core electrons from the Schrödinger equation, thus making lead (Z = 82) not greatly more difficult to treat than carbon (Z = 6). Much of the early work on pseudopotentials is indebted in one way or another to that of Phillips and Kleinman,[a] whose primary interest was in solid state physics, where *ab initio* treatments remain impractical except for the very simplest systems. For molecules one must be more precise, since the *ab initio* result one is attempting to model is available (at least for smaller test cases) for immediate comparison. Most electronic structure theorists would agree that the key contribution to the development of a more rigorous pseudopotential was the paper by Kahn and Goddard. Although significant improvements have since been reported,[b] there is a considerable flavour of the Kahn–Goddard approach in most contemporary molecular pseudopotentials.

[a] J. C. Phillips and L. Kleinman, New method for calculating wavefunctions in crystals and molecules, *Phys. Rev.* **116**, 287 (1959).

[b] P. A. Christiansen, Y. S. Lee, and K. S. Pitzer, Improved *ab initio* effective core potentials for molecular calculations, *J. Chem. Phys.* **71**, 4445 (1979).

98

W. J. HUNT, P. J. HAY, & W. A. GODDARD

Self-consistent procedures for generalized valence bond wave functions. Applications H_3, BH, H_2O, C_2H_6, and O_2

J. Chem. Phys. **57**, 738 (1972)

The problem with the SOGI or spin-optimized SCF methods of Goddard, Kaldor, and Harris was their use of non-orthogonal orbitals. The general evaluation of hamiltonian matrix elements in terms of such a basis involves $n!$ (n is the number of electrons) contributions and thus to date no more than eight electrons have been explicitly treated in this manner. Here Hunt, Hay, and Goddard showed that a particular type of multiconfiguration SCF wavefunction gives a qualitatively reasonable approximation to the $G1$ (or maximally paired SCF) wavefunction. Although the two orbitals singlet spin coupled to describe a particular pair of electrons need not be orthogonal to each other, they are required to remain orthogonal to the molecular orbitals for the other electron pairs. Thus this generalized valence bond (GVB) model is a special case of the separated pair or strongly orthogonal geminal wavefunction. The genius of the GVB method lies in the fact that the resulting energy expression involves only integrals of the type h_i, J_{ij}, and K_{ij}. Therefore, such GVB wavefunctions may be determined via relatively straightforward SCF procedures. The Cal Tech group has successfully applied these methods to a broad spectrum of chemical problems.

99

E. CLEMENTI & H. POPKIE

Study of the structure of molecular complexes. I. Energy surface of a water molecule in the field of a lithium positive ion

J. Chem. Phys. **57**, 1077 (1972)

For our purposes the primary significance of this paper lies not in the material described by the title. Of more importance from a purely theoretical perspective is the Appendix, titled 'Choice of the basis set for the water molecule'. Using van Duijneveldt's vigorously optimized sp gaussian functions for the O and H atoms, Clementi and Popkie reported SCF energies from 68 different uncontracted basis sets for H_2O at its equilibrium geometry. If one takes the O(9s 5p), H(5s) set as a starting point, it may be seen for example that increasing the oxygen set to O(13s 8p) lowers the SCF energy by 0.0102 hartree. In addition 26 basis sets including polarization functions were also evaluated. The latter compendium shows for example that after a large O(spd), H(sp) basis set is chosen the addition of f functions on oxygen and d functions on hydrogen lowers the SCF energy by only 0.0013 hartree. The reader is referred elsewhere for systematic studies of ethylene and acetylene in the same spirit, but using necessarily smaller basis sets.[a]

[a] C. W. Bock, P. George, G. J. Mains, and M. Trachtman, An *ab initio* study of the dependence of molecular geometry on basis set. Part I. Ethylene, *J. Mol. Struct.* **49**, 215 (1978); Part II. Acetylene, *J. Mol. Struct.* **51**, 307 (1979).

100

C. F. BENDER, S. V. O'NEIL,
P. K. PEARSON &
H. F. SCHAEFER

Potential energy surface including electron correlation for $F + H_2 \rightarrow FH + H$: refined linear surface

Science **176**, 1412 (1972)

At the time of its publication, this paper was of general interest as the first serious attempt to semi-quantitatively describe an A + BC energy surface for a system larger than the classic H_3. The manuscript was to a significant degree successful in that regard. In retrospect, however, the long-term theoretical impact was rather in the discovery that the predicted SCF barrier height for such reactions is in poor agreement with the experimental activation energy. The explicit treatment of electron correlation is absolutely essential for even a qualitative description of transition states of this type. Using a double zeta plus polarization (DZ + P) basis set the predicted SCF barrier height was 29.3 kcal, compared with the CI result of 1.66 kcal, which is essentially coincident with the experimental activation energy of ~1.7 kcal. Moreover, the transition state geometry predicted at the SCF level [r(F − H) = 1.18 Å, r(H − H) = 0.84 Å] differed greatly from the correlated result [r(F − H) = 1.54 Å, r(H − H) = 0.77 Å]. The finding that transition states are more sensitive to correlation effects than are equilibrium structures has since been confirmed for a large number of other molecular systems.[a]

[a] See for example, H. F. Schaefer, Interaction potentials I: atom–molecule potentials, pages 45–78 of *Atom-molecule collision theory*, edited by R. B. Bernstein (Plenum, New York, 1979).

101

K. RUEDENBERG, R. C. RAFFENETTI, & R. D. BARDO

Even tempered orbital bases for atoms and molecules

pages 164–9 of *Energy, Structure, and Reactivity*,
editors D. W. Smith and W. B. McRae (Wiley, New York, 1973).

This paper is difficult to locate, appearing as a five-page long contributed remark in the Proceedings of the Summer 1972 Conference on Theoretical Chemistry, held at the University of Colorado. Nevertheless this is the first public announcement of the development by the Iowa State group of the concept of even-tempered basis sets. In an even-tempered basis set all members have the same overlap integral with their immediately adjacent (in terms of spatial extent) partners of the same atomic symmetry. The outcome of this requirement is that any two successive orbital exponents of a sequence of even-tempered functions have the same ratio. Thus the complete specification of a particular set of even-tempered functions (e.g., eleven 1s gaussian functions for the carbon atom) requires only the starting orbital exponent and the (constant) ratio between successive exponents. Ruedenberg and coworkers noted that where the exhaustive optimization of atomic basis sets was reported in the literature, the even-tempered criterion was almost always approximately satisfied. The method thus greatly simplifies the optimization of atomic basis sets, while providing an optimally distributed basis set in the sense of uniform overlap.

102

J. ROSE, T. SHIBUYA, & V. McKOY

Application of the equations-of-motion method to the excited states of N_2, CO, and C_2H_4

J. Chem. Phys. **58**, 74 (1973)

The work of Kelly and others on many-body perturbation theory created much interest on the part of chemical theorists in non-variational methods. Some of the earliest and most far-reaching of these studies were carried out by McKoy and his co-workers at Cal Tech. The present paper presents an impressive snapshot of the equations-of-motion (EOM) method[a] about three years into its molecular development. A great virtue of this method is that it provides electronic excitation energies directly as opposed to solving Schrödinger's equation separately for the initial and final state energies. Equally important the intensities of the different transitions (i.e., transition probabilities) fall out of this formalism in a simple way. In the present study, the predicted excitation frequencies of nine states of CO and eleven states of N_2 were all found to lie within 10 per cent of the available experimental energy differences. This was accomplished using single particle-hole and two particle-hole components of the excitation operators, and may be viewed as an alternative to CI involving single and double excitations relative to the closed-shell ground state Hartree–Fock reference function.

[a] For a formal review of the EOM method, see D. J. Rowe, Equations-of-motion method and the extended shell model, *Rev. Mod. Phys.* **40**, 153 (1968).

103

W. MEYER

PNO–CI studies of electron correlation effects. I. Configuration expansion by means of nonorthogonal orbitals and application to the ground state and ionized states of methane

J. Chem. Phys. **58**, 1017 (1973)

Meyer's pioneering 1971 paper on the PNO–CI method provided a primarily intuitive description of that method. The second paper presents the mathematical foundations of the method, as well as extensive applications to CH_4 and CH_4^+. It seems indisputable that the PNO–CI method provides the most rapidly convergent (fewest numbers of configurations → largest fraction of correlation energy) CI expansion generally available at the present time. However, it should be noted that the method yields an energy somewhat higher than straightforward CISD with the same basis set. This is because the PNO–CI method employs a truncated set of pair natural orbitals, typically yielding ~10 configurations, to describe each electron pair.

Also elaborated on in this paper is the coupled electron pair approximation (CEPA) introduced in the earlier Meyer paper under the designation cluster-corrected CI. CEPA provides a nonvariational estimate of the importance of a certain class of higher excitations, namely unlinked clusters. This is accomplished by solving a second eigenvalue problem in which the PNO–CI hamiltonian matrix has its diagonal elements shifted in a particular fashion.[a]

[a] For a recent discussion, see R. Ahlrichs and C. Zirz, CEPA model and MBPT, in *Quantum Chemistry into the 1980s*, edited by P. G. Burton, Proceedings of the Molecular Physics and Quantum Chemistry Workshop, Wollongong, Australia, February, 1980.

104

R. C. RAFFENETTI

General contraction of gaussian atomic orbitals: core, valence, polarization, and diffuse basis sets; molecular integral evaluation

J. Chem. Phys. **58**, 4452 (1973)

Prior to Raffenetti's paper, essentially all contracted gaussian basis sets could be described as *segmented*. That is, each primitive function occurred in only a single contracted function. Quite simply, this was because the occurrence of the same primitive gaussian function in two contracted functions was equivalent to adding a primitive gaussian to the effort required for two-electron integral evaluation. This practical restriction to segmented basis sets is a serious one because several primitive basis functions typically contribute, for example to the 1s and 2s Hartree–Fock atomic orbitals. These 'overlapping' primitives must be left totally uncontracted in a segmented scheme to avoid significant losses in the total energy. Raffenetti's insight was to compute clusters of related integrals over primitive gaussian basis functions and then transform them to integrals over arbitrarily contracted functions. In this way the full atomic flexibility of the gaussian basis set may be retained, while the total number of contracted functions remains modest. The general contraction scheme has not yet been universally applied, but presumably will be when other innovations in the computation of two-electron integrals run their course.

105

G. C. LIE, J. HINZE, & B. LIU

Valence excited states of CH. I. Potential curves

J. Chem. Phys. **59**, 1872 (1973)

After Wahl and Das, the principal contributor to MCSCF theory during the late 1960s and early 1970s was Hinze. In particular his 1973 paper[a] presented a number of insights which continue to be of value. Moreover, Hinze developed a general MCSCF program that was widely utilised, particularly by his collaborators at IBM San Jose. One of the most important studies issuing from the new MCSCF method was the present paper on the five lowest electronic states of CH. The paper also established the pattern of excellence now expected from the IBM group in studies of electron correlation in small molecules.[b] For example, a very large Slater orbital basis set was chosen: C(6s 4p 2d 2f), H(4s 3p 2d). The MCSCF wavefunctions included all configurations (up to eight) required to properly describe the appropriate asymptotic limits C(^{3}P, ^{1}D, or ^{1}S) + H(^{2}S). CISD relative to this multi-configuration reference function yielded as many as 4147 configurations for the final calculations. As expected, close agreement (typically 0.002 Å in bond distances and 0.1 eV in dissociation energies) with experiment was found.

[a] J. Hinze, MC-SCF. I. The multi-configuration self-consistent-field method, *J. Chem. Phys.* **59**, 6424 (1973).

[b] For a general discussion of the multi-reference CI philosophy, see P. S. Bagus, B. Liu, A. D. McLean, and M. Yoshimine, Application of wave mechanics to the electronic structure of molecules through configuration interaction, pages 99–118 of *Wave mechanics: the first fifty years*, edited by W. C. Price, S. S. Chissick, and T. Ravensdale (Butterworth, London, 1973).

106

R. M. PITZER

Electron repulsion integrals and symmetry adapted charge distributions

J. Chem. Phys. **59**, 3308 (1973)

For systems of high symmetry, the solution of the Hartree–Fock equations can often be greatly simplified relative to a straightforward approach. Such simplifications are masterfully illustrated for atoms by the definitions of unique integrals appearing in the 1963 Roothaan–Bagus review. The paper cited here and related work[a] by Pitzer established how to enumerate the minimum list of SCF-required integrals for non-linear polyatomic molecules of high symmetry. The essential idea is that two-electron integrals over symmetry adapted basis functions may be obtained without the evaluation of all the constituent integrals over basis functions. An example of the importance of these ideas is provided by a subsequent study[b] of the permanganate ion MnO_4^- using a large basis of size Mn(16s 13p 6d/10s 7p 3d), O(10s 6p/5s 3p). This same system had previously been investigated by Johansen[c] who computed *ca.* 4 000 000 P supermatrix elements (Roothaan's nomenclature), while the new method of Pitzer reduced this number to 117 006.

[a] R. M. Pitzer, Contribution of atomic orbital integrals to symmetry orbital integrals, *J. Chem. Phys.* **58**, 3111 (1973).
[b] H. Hsu, C. Peterson, and R. M. Pitzer, Calculations on the permanganate ion in the ground and excited states, *J. Chem. Phys.* **64**, 791 (1976).
[c] H. Johansen, SCF LCAO MO calculation for MnO_4^-, *Chem. Phys. Lett.* **17**, 569 (1972).

107

I. SHAVITT, C. F. BENDER, A. PIPANO, & R. P. HOSTENY

The iterative calculation of several of the lowest or highest eigenvalues and corresponding eigenvectors of very large symmetric matrices

J. Comput. Phys. **11**, 90 (1973)

Nesbet's method for extracting the lowest eigenvalue of a large CI matrix is an iterative procedure in which one component at a time of a trial vector is adjusted so as to satisfy the corresponding equation of the eigenvalue problem. For some years prior to 1973, Shavitt and his colleagues had been working to improve upon and extend the applicability of Nesbet's method, and the present paper describes their approach, a modified method of optimal relaxation (MOR). The Nesbet algorithm appears as a limiting case of the more general MOR, which generally provided more rapid convergence when the initial trial vector was a poor approximation to the eventual solution. More importantly, the Shavitt method was readily applied to higher-lying energy eigenvalues as well. Since one often wishes to determine the energies of two or three electronic states of the same space-spin symmetry (e.g., $^1A'$), this capability is required for general CI methods. For especially difficult cases (for example, those involving nearly degenerate roots), an extrapolation procedure was developed to accelerate convergence.

108

M. YOSHIMINE

Construction of the hamiltonian matrix in large configuration interaction calculations

J. Comput. Phys. **11**, 449 (1973)

In conventional CI methods, when one constructs the hamiltonian matrix, there are three long data lists which must be simultaneously processed. These are the two-electron integrals, the hamiltonian matrix elements H_{ij}, and the coefficients which tell how each two-electron integral appears in each H_{ij}. The third of these lists is often referred to as the 'formula tape' and is independent of molecular geometry if, for example, one restricts consideration to a particular chemical reaction. Typically each of these three lists is too long to be stored entirely in the central memory of a modern digital computer. This paper introduced the well-known 'Yoshimine sort' for the simultaneous processing of several long data lists. Yoshimine showed that this could be done very efficiently using direct (or random) access to files stored on disc drives of various types. This reordering of long lists is also necessary for several other quantum mechanical methods and has become a standard procedure among *ab initio* theorists. In newly developing techniques, such as those for the evaluation of CI analytic second derviatives, the Yoshimine sort becomes indispensable.

109

U. KALDOR

Many-body perturbation-theory calculations with finite, bound basis sets

Phys. Rev. A **7**, 427 (1973)

Perhaps surprisingly, the title of this paper describes exactly what Kaldor accomplished. We have noted Schulman and Kaufman's work in this respect for H_2, but they considered only diagrams of low order and had no basis for comparison with the standard atomic MBPT based on a complete set of bound and continuum numerical orbitals. Kaldor chose the beryllium atom, the topic of Kelly's two earliest papers, and used a large analytic (Slater function) basis set of size 10s 8p 6d 4f. This basis set gave Kaldor energetic convergence for individual diagrams to better than 0.0001 hartree. All second- and third-order diagrams were precisely evaluated, and agreement with Kelly's numerical results was very good except where the latter made certain approximations. Kaldor's entire MBPT treatment recovered 97.8 per cent of the correlation energy, very close to the 97.1 per cent obtained variationally by Bunge in 1968. This paper cleared the way for widespread applications of MBPT to molecular systems, where analytic basis sets were mandatory.

110

M.-M. COUTIERE, J. DEMUYNCK, & A. VEILLARD

Ionization potentials of ferrocene and Koopmans' theorem An *ab initio* LCAO–MO–SCF calculation

Theoret. Chim. Acta **27**, 281 (1972)

The 1965 paper of Bagus established the degree to which Hartree–Fock ionization potentials for closed-shell atoms may be expected to agree with experiment. In general both Koopmans' theorem and ΔE_{SCF} (the difference between the Hartree–Fock energies of the atom and its positive ion) give qualitative agreement, with quantitative agreement often found for the latter method. The paper by Coutiere, Demuynck, and Veillard established that for at least one class of molecules (organometallics), one should not even expect qualitatively reasonable results from Koopmans' theorem. Specifically, Veillard noted the enormity of the relaxation energy for transition metal atom d shells. For molecular orbitals dominated by ligand character, there is little electronic rearrangement upon ionization, and hence the ΔE_{SCF} ionization potential is close to the corresponding orbital energy. However for the metal 3d-like orbitals there is a marked rearrangement upon ionization. While these ferrocene orbitals include a small amount of ligand orbital character for the neutral molecule, they become nearly pure metal 3d orbitals for the ion, and Veillard reported differences as large as 6 eV between the ΔE_{SCF} and Koopmans' ionization potentials. As a result the ordering of the orbital energies is not likely to agree with the observed photoelectron spectra. Later studies of ferrocene[a] and many other organotransition metal species have confirmed the generality of Veillard's finding.

[a] P. S. Bagus, U. I. Walgren, and J. Almlöf, A theoretical study of the electronic structure of ferrocene and ferricinium: application to Mössbauer isomer shifts, ionization potentials, and conformation, *J. Chem. Phys.* **64**, 2324 (1976).

111

S. R. LANGHOFF & E. R. DAVIDSON

Configuration interaction calculations on the nitrogen molecule

Int. J. Quantum Chem. **8**, 61 (1974)

We have already mentioned Meyer's CEPA method for the estimation of the importance of that particular class of higher (than double) excitations known as unlinked clusters. Here an even simpler method for the energetic importance of quadruple excitations was proposed,[a] namely

$$\Delta E_{\mathrm{Q}} = (1 - C_0^2)\Delta E_{\mathrm{D}}$$

in which ΔE_{D} is the correlation energy due to double excitations and C_0 is the coefficient of the Hartree–Fock reference function in the CI wavefunction including all double excitations. Langhoff and Davidson showed that this equation (thereafter known as the 'Davidson correction') gives a fairly reliable estimate of the contribution of quadruple excitations, after the dominant double excitations have been treated variationally. Formally this correction approximately represents the unlinked fourth-order (in many-body perturbation theory) contribution of quadruple excitations which must be cancelled (in MBPT) by renormalization contributions from double excitations.[b] This simple formula has proven qualitatively successful in a large number of subsequent applications.

[a] This formula is also given by E. R. Davidson in the article Configuration interaction description of electron correlation, pages 17–30 of *The world of quantum chemistry*, editors R. Daudel and B. Pullman (D. Reidel, Dordrecht, Holland, 1974).
[b] E. R. Davidson and D. W. Silver, Size consistency in the dilute helium gas electronic structure, *Chem. Phys. Lett.* **52**, 403 (1977).

112

P. J. HAY & I. SHAVITT

Ab initio configuration interaction studies of the π-electron states of benzene

J. Chem. Phys. **60**, 2865 (1974)

By 1974 Shavitt was recognized as one of the few theoretical chemists with a truly deep understanding of the correlation problem. This paper reflects many of the insights he had accumulated since the years of graduate study with Boys. At the time, of course, benzene was a rather large molecule to be treated via CI with a respectable basis set. The Huzinaga–Dunning double zeta set was augmented here with two diffuse π functions on each carbon atom, for a total of 84 contracted functions. The four-index transformation of such a large basis set was accomplished by exploiting the fact that the SCF σ orbitals did not participate in the CI. Thus it was possible to define an effective potential, with which the σ (or frozen core) electrons do not appear explicitly in the CI.[a] Full use of the degenerate D_{6h} symmetry was made in the CI, which was very extensive within the π electron space, including as many as 2636 symmetry- and spin-adapted configurations for the $^3E_{1u}$ state. Vertical excitation energies were found in very good agreement with available experimental data, except for those mixed states analogous to the earlier discussed $\pi \rightarrow \pi^*$ singlet state of ethylene.

[a] This general approach was first used by T. H. Dunning, R. P. Hosteny, and I. Shavitt, Low-lying π-electron states of *trans*-butadiene, *J. Amer. Chem. Soc.* **95**, 5067 (1973).

113

J. PALDUS

Group theoretical approach to the configuration interaction and perturbation theory calculations for atomic and molecular systems

J. Chem. Phys. **61**, 5321 (1974)

This is perhaps the most significant of a series of important papers[a] in which Paldus introduced the unitary group approach (UGA) to theoretical chemistry. As early as 1950 the Russian workers Gelfand and Tsetlin had shown how to use the unitary group $U(n)$ on $n \times n$ matrices to construct an orthonormal basis and to evaluate matrix elements of the generators of this basis. However the relevance of the unitary group approach to the many-body problem was first seized upon by the nuclear physicist Moshinsky[b] in the 1960s. Nevertheless, it is probably fair to say that the developments ushered in by Paldus's formulation of the UGA for electronic systems have already been of greater importance for chemistry than were the earlier advances for physics. Specifically, Paldus showed that the restriction to many-electron systems allows a great simplification of the general structure of the UGA. The determination of electronic hamiltonian matrix elements was shown to follow in a straightforward manner from the matrix representatives of the generators E_{ij} in the Gelfand-Tsetlin basis.

[a] A summary of the early work is given in J. Paldus, Many-electron correlation problem. A group theoretical approach, pages 131-290 of Volume 2, *Theoretical chemistry: advances and perspectives,* editors H. Eyring and D. J. Henderson (Academic Press, New York, 1976).

[b] M. Moshinsky, *Group theory and the many-body problem* (Gordon and Breach, New York, 1968).

114

F. SASAKI & M. YOSHIMINE

Configuration-interaction study of atoms. I. Correlation energies of B, C, N, O, F, and Ne

Phys. Rev. A **9**, 17 (1974)

Even though this paper is now more than nine years old, it remains the definitive systematic theoretical study of correlation effects in first row atoms. The very large Slater orbital sets used in the CI treatments included 8s, 7p, 6d, 5f, 4g, 3h, and 2i orthogonal functions. The lists of configurations employed included all single and double excitations relative to the appropriate Hartree-Fock reference function, amounting to as many as 1571 LS eigenfunctions for the fluorine atom. A smaller number (necessarily very carefully selected) of triple and quadruple excitations were added, to yield final CI expansions ranging in size from 798 (boron) to 2649 (fluorine) configurations. This consistent study provided variational predictions between 95 and 97 per cent of the experimental correlation energies. The energy effects of triple and higher excitations were estimated to be of the order of 3-4 per cent of correlation energy for these systems containing between five and ten electrons. A second paper[a] in the series showed that even at this very sophisticated level of theory, predicted electron affinities may be as much as 0.3 eV less than experiment.

[a] F. Sasaki and M. Yoshimine, Configuration-interaction study of atoms. II. Electron affinities of B, C, N, O, and F, *Phys. Rev. A* **9**, 26 (1974).

115

R. AHLRICHS, H. LISCHKA, V. STAEMMLER, & W. KUTZELNIGG

PNO–CI (pair natural orbital configuration interaction) and CEPA–PNO (coupled electron pair approximation with pair natural orbitals) calculations of molecular systems. I. Outline of the method for closed-shell states

J. Chem. Phys. **62**, 1225 (1975)

Since Kutzelnigg's group had championed the use of pair natural orbitals (PNOs) within the framework of the IEPA, it was by no means surprising that they would take an intense interest in Meyer's newly developed PNO-CI method. Although the conceptual basis of the present paper is largely the work of Meyer, the actual formulation of the method presented here is somewhat different and in several respects clearer than in Meyer's papers. Also included is a lucid exposition of the CEPA method for the estimation of the importance of unlinked clusters. The new implementation of PNO-CI and CEPA was used straight away in a variety of chemical applications and quickly became one of the standard methods for the description of electron correlation in molecules. One of the bolder applications[a] of the method was to the nitrogen molecule, where up to *sextuple* excitations are required for dissociation to two Hartree-Fock ground state N atoms.[b] Perhaps surprisingly, the CEPA potential energy curve is quite accurate out to bond distances about 50 per cent greater than the equilibrium value.

[a] R. Ahlrichs, H. Lischka, B. Zurawski, and W. Kutzelnigg, PNO–CI and CEPA–PNO calculations of molecular systems. IV. The molecules N_2, F_2, C_2H_2, C_2H_4, and C_2H_6, *J. Chem. Phys.* **63**, 4685 (1975).

[b] G. C. Lie and E. Clementi, Correlation energy corrections as a functional of the Hartree–Fock type density and its application to the homonuclear diatomic molecules of the second row atoms, *J. Chem. Phys.* **60**, 1288 (1974).

116

E. A. McCULLOUGH

The partial-wave self-consistent-field method for diatomic molecules: computational formalism and results for small molecules

J. Chem. Phys. **62**, 3991 (1975)

In the best of all possible worlds, one would surely want to do away with analytic basis functions. All molecular orbitals would be represented in a strictly numerical form, following Hartree, and determined directly via general self-consistent-field procedures. Just think of the relief associated with never again having to be concerned about the perilous effects of extending one's finite basis set.[a] With the work of McCullough in 1975, this wish came very much closer to being a reality. For light diatomic molecules, McCullough showed using the elliptical coordinates ξ and η that it is possible to determine the numerical Hartree-Fock wavefunctions to a high degree of precision. In the present and subsequent papers,[b] it was demonstrated that this approach provides a testing ground for energies and other properties (such as polarizabilities) determined using the more conventional Slater and gaussian basis sets. Quite recently the method has been successfully extended[c] to diatomic MCSCF wavefunctions.

[a] A more cautious reviewer has suggested that numerical techniques only serve to replace basis set concerns with anxiety about the adequacy of the number and distribution of quadrature points.

[b] P. A. Christiansen and E. A. McCullough, Numerical Hartree–Fock calculations for N_2, FH, and CO: comparison with optimized LCAO results, *J. Chem. Phys.* **67**, 1877 (1977).

[c] L. Adamowicz and E. A. McCullough, A numerical multiconfiguration self-consistent-field method for diatomic molecules, *J. Chem. Phys.* **75**, 2475 (1981).

117

C. W. BAUSCHLICHER, D. H. LISKOW, C. F. BENDER, & H. F. SCHAEFER

Model studies of chemisorption. Interaction between atomic hydrogen and beryllium clusters

J. Chem. Phys. **62**, 4815 (1975)

In retrospect, the 1970s may be recognized as the decade in which relatively large numbers of chemists began to take a serious interest in surface chemistry. The paper in question represents the first systematic *ab initio* study of chemisorption, dealing with the interaction between monatomic hydrogen and the (0001) surface of metallic beryllium. Perhaps surprisingly, a good deal is known experimentally about this particular metal surface, including the fact that it is not reconstructed, i.e., the surface atoms retain their approximate relative positions from the bulk. Beginning with Be_3, finite clusters were used to model the surface up to Be_{10} (taken to have seven surface and three second layer atoms), which turns out to provide a satisfactory model for chemisorption. The general conclusion of this work and later *ab initio* studies (involving clusters as large as Be_{36})[a] was that a qualitative description of chemisorption results when the cluster model includes all metal atoms adjacent to the adsorbed species, and in addition all the nearest neighbours of the above specified metal atoms.

[a] P. S. Bagus, H. F. Schaefer, and C. W. Bauschlicher, The convergence of the cluster model for the study of chemisorption: Be_{36} H, *J. Chem. Phys.* **78**, 1390 (1983).

118

E. R. DAVIDSON

The iterative calculation of a few of the lowest eigenvalues and corresponding eigenvectors of large real-symmetric matrices

J. Comput. Phys. **17**, 87 (1975)

For a problem as central as the determination of eigenvalues, one intuitively expects to see gradual progress made from year to year. However Davidson's contribution of 1975 stands as the method of choice more than eight years after its publication. Hence this paper is required reading for new graduate students in electronic structure theory. The method may be viewed as a compromise between Nesbet and Shavitt's ideas and the tridiagonalization scheme introduced much earlier by Lanczos.[a] The primary advantage of Davidson's method relative to the earlier discussed MOR procedure is accelerated convergence. In the specific cases studied (for example the four lowest $^1\Sigma^+$ electronic states of LiF)[b] the MOR method occasionally required more than 100 iterations, while the new method converged in no more than 20 iterations. In addition Davidson's method is not adversely affected by nearly degenerate eigenvalues, requires core storage of only two eigenvectors at once, and can in principle be used to locate higher eigenvalues directly, without finding accurate values of the lower roots first.

[a] C. Lanczos, An iteration method for the solution of the eigenvalue problem of linear differential and integral operators, *J. Res. Nat. Bur. Stand.* **45**, 255 (1950).

[b] L. R. Kahn, P. J. Hay, and I. Shavitt, Theoretical study of curve crossing: *ab initio* calculations on the four lowest $^1\Sigma^+$ states of LiF, *J. Chem. Phys.* **61**, 3530 (1974).

119

G. H. F. DIERCKSEN, W. P. KRAEMER, & B. O. ROOS

SCF–CI studies of correlation effects on hydrogen bonding and ion hydration. The systems: H_2O, $H^+ \cdot H_2O$, $Li^+ \cdot H_2O$, $F^- \cdot H_2O$, and $H_2O \cdot H_2O$

Theoret. Chim. Acta **36**, 249 (1975)

Perhaps the most impressive of the early applications of Roos's direct CI method is presented here. Since the water dimer is of such fundamental importance in physical chemistry, the study was very timely. Their contracted gaussian basis of size O(5s 4p 1d), H(3s 1p) meant that a total of 66 molecular orbitals were included for $(H_2O)_2$. Straightforward valence shell CISD was carried out via the direct CI approach, involving 56 268 $^1A'$ configurations, an astonishingly large number at the time. Eleven such calculations were performed to determine the relative orientation of the two rigid water molecules within the dimer, and a structure rather close to the SCF geometry was found. Correlation effects were found to increase the predicted dimerization energy D_e from 5.1 kcal to 6.1 kcal, but estimates of the zero-point vibrational energies suggested that D_e may be ~1 kcal greater than D_0, the dissociation energy which would be observed in a definitive experiment. No such experiment yet exists but a subsequent theoretical study by Clementi and co-workers[a] provided qualitative support for the Diercksen study.

[a] O. Matsuoka, E. Clementi, and M. Yoshimine, CI study of the water dimer potential surface, *J. Chem. Phys.* **64**, 1351 (1976).

120

R. AHLRICHS & F. DRIESSLER

Direct determination of pair natural orbitals. A new method to solve the multi-configuration Hartree–Fock problem for two-electron wavefunctions

Theoret. Chim. Acta **36**, 275 (1975)

A primary distinction between Kutzelnigg's formulation of PNO-CI and the original Meyer approach was in the method used for the determination of the approximate pair natural orbitals. That innovation is the topic of the cited paper of Ahlrichs and Driessler. For the limiting case of a two-electron system, the Ahlrichs-Driessler method, an iterative Hartree-Fock like method, solves the Schrödinger equation exactly for the particular finite basis set adopted. Moreover, the new method does this while requiring 'less computational work than a conventional Hartree-Fock computation within the same basis set'. To make things even better, the method simultaneously yields the CI coefficients and the total energy for the natural orbital two-electron wavefunction, rendering unnecessary a final CI calculation. For many-electron systems, the approach of Ahlrichs and Driessler was about an order of magnitude more efficient than earlier methods. In practice this meant that the fraction of computation going into the determination of PNOs became essentially negligible in the overall scheme of things. Finally, and perhaps most important, this method led Meyer to the introduction one year later of the method of self-consistent electron pairs.

121

H. F. SCHAEFER & W. H. MILLER

Large scale scientific computation via minicomputer

Computers and Chemistry **1**, 85 (1976)

Prior to 1972 it had been almost universally assumed that state-of-the-art *ab initio* quantum mechanics was only possible on state-of-the-art large-scale computers. The fact that a notable fraction of the pioneering work in the area was done by scientists at IBM tended to confirm this view. However, in the early 1970s the failure of the CDC Star supercomputer, the abandonment of large scale computer development by IBM, and the rapid development of minicomputers led the authors of this paper to a different conclusion. In February of 1973 they proposed to the U.S. National Science Foundation the use of a minicomputer for theoretical chemistry. The machine arrived in November of the same year and proved itself to be within a factor of 30 in speed of the fastest existing large scale computer, the CDC 7600. One decade later, the distribution of small and large computers is quite different, with many advanced theoretical groups relying exclusively on minicomputers for their research. Further documentation of the experience with the first theoretical chemistry minicomputer is provided elsewhere.[a]

[a] P. K. Pearson, R. R. Lucchese, W. H. Miller, and H. F. Schaefer, Theoretical chemistry via minicomputer, pages 171–190 of *Minicomputers and large scale computations*, edited by P. Lykos (American Chemical Society Symposium Series 57, Washington, D.C., 1977).

122

S. R. LANGHOFF & E. R. DAVIDSON

Ab initio evaluation of the fine structure and radiative lifetime of the $^3A_2$ ($n \rightarrow \pi^*$) state of formaldehyde

J. Chem. Phys. **64**, 4699 (1976)

The incorporation into the hamiltonian of terms other than the ordinary electrostatic contributions—kinetic energy, electron-nuclear attraction, and electron repulsion—has long been considered a challenge to electronic structure theorists. Specifically, the treatment of the spin-orbit interaction is a particularly treacherous problem. Since this term is essential to an understanding of triplet state fine structure and triplet–singlet decay rates, the problem is central to a number of general questions in photophysics and photochemistry.[a] Perhaps the first satisfactory treatment of spin-orbit coupling in any polyatomic molecule was the work of Langhoff and Davidson on formaldehyde. The complete microscopic spin-orbit hamiltonian was employed to evaluate matrix elements connecting the $^3A_2$ (n → π^*) state CI wave-function with those for twelve low-lying states of H_2CO. These matrix elements were then used to determine radiative lifetimes for the three $^3A_2$ sublevels T_x, T_y, and T_z, among which energy differences are governed by the zero-field splitting parameters D and E. The theoretical predictions were in rough but satisfactory agreement with available experiments.

[a] A number of interesting astrophysical predictions have also been made recently on the basis of *ab initio* studies of spin-orbit coupling. See, for example, W. G. Richards, H. P. Trivedi, and D. L. Cooper, *Spin-orbit coupling in molecules* (Clarendon Press, Oxford, 1981).

123

M. DUPUIS, J. RYS, & H. F. KING

Evaluation of molecular integrals over Gaussian basis functions

J. Chem. Phys. **65**, 111 (1976)

For gaussian sp basis sets, the most effective method for the evaluation of two-electron integrals remains that introduced by the Pople group[a] in the GAUSSIAN-70 program. However as soon as d functions are introduced, it is now generally accepted that the Rys polynomial procedure, introduced by King's group at Buffalo, becomes the method of choice. Briefly, they showed that any electron repulsion integral over primitive gaussian functions may be expressed as a finite sum of products of three factors I_x, I_y, and I_z corresponding to the three cartesian coordinates. Each term corresponds to one root of a Rys polynomial,[b] so named after the mathematical analysis of one of King's graduate students. The Rys polynomial scheme was designed to make excellent use of information common to entire 'shell blocks' of integrals, such as (SP|DF), taken to designate the collection of 180 different integrals, one of which would be $(sp_x | d_{xy}\ f_{xyz})$. These methods were incorporated in the HONDO program, which quickly gained widespread approval within the theoretical community.

[a] J. A. Pople and W. J. Hehre, Computation of electron repulsion integrals involving contracted gaussian basis functions, *J. Comput. Phys.* **27**, 161 (1978).
[b] H. F. King and M. Dupuis, Numerical integration using Rys polynomials, *J. Comput. Phys.* **21**, 144 (1976).

124

P. ROSMUS & W. MEYER

Spectroscopic constants and the dipole moment functions for the $^1\Sigma^+$ ground state of NaLi

J. Chem. Phys. **65**, 492 (1976)

As noted earlier, Wahl's predictions of the bond energy and vibrational frequency for NaLi were beautifully confirmed when the molecule was subsequently observed spectroscopically. However the theoretical dipole moment, μ = 1.24 debye, proved far from the mark when Dagdigian and Wharton's molecular beam experiment[a] yielded μ = 0.46 ± 0.01 debye. A second ostensibly reliable theoretical treatment[b] predicted μ = 0.99 debye, also in unacceptable agreement with experiment. The discrepancy was resolved in this paper by Rosmus and Meyer. Using a large gaussian basis set, these authors first established the near-Hartree–Fock value μ = 0.72 debye. However, valence shell correlation effects (which involve only the two electrons forming the NaLi single bond) were found to increase μ to 0.95 debye. Rather unexpectedly, Rosmus and Meyer then discovered that core-valence correlation effects significantly reduced the predicted dipole moment, to 0.49 debye. The latter result is in good agreement with experiment and demonstrated forcefully that core-valence considerations can be very important for systems with only a few valence electrons.

[a] J. Graff, P. J. Dagdigian, and L. Wharton, Electric resonance spectrum of NaLi, *J. Chem. Phys.* **57**, 710 (1972).
[b] S. Green, Electric dipole moment of diatomic molecules by configuration interaction. I. Closed shell molecules, *J. Chem. Phys.* **54**, 827 (1971).

125

C. E. DYKSTRA, H. F. SCHAEFER, & W. MEYER

A theory of self-consistent electron pairs. Computational methods and preliminary applications

J. Chem. Phys. **65**, 2740 (1976)

In the two-electron method of Ahlrichs and Driessler, all matrix elements relevant to the iterative process are obtained from a single exchange operator. This was generalized to many-electron systems by Meyer in the theory of self-consistent electron pairs (SCEP).[a] Meyer's achievement was by no means obvious, since the single exchange operator required for the two electron case is replaced in SCEP by $N(N+1)/2$ internal pair coulomb and exchange operators (N = number of doubly occupied SCF orbitals) plus N^2 external exchange operators. Within one year, the initial development of the new method was completed via collaboration between the Berkeley and Mainz research groups. SCEP is qualitatively different from other variational methods in that the wavefunction is not expressed in terms of configurations and CI coefficients. Instead the correlation function for each electron pair is expressed as a coefficient matrix of size $m \times m$, where m is the number of basis functions. SCEP is however equivalent to a CI involving all double excitations, and single excitations may easily be appended in a variational manner. Over the past several years Dykstra[b] has continued to develop the method, which is now competitive with sophisticated CI techniques.[c]

[a] W. Meyer, Theory of self-consistent electron pairs. An iterative method for correlated many-electron wavefunctions, *J. Chem. Phys.* **64**, 2901 (1976).

[b] C. E. Dykstra, R. A. Chiles, and M. D. Garrett, Recent computational developments with the self-consistent electron pairs method and application to the stability of glycine conformers, *J. Comput. Chem.* **2**, 266 (1981).

[c] See also H. -J. Werner and E. -A. Reinsch, The self-consistent electron pairs method for multiconfiguration reference state functions, *J. Chem. Phys.* **76**, 3144 (1982).

126

L. G. YAFFE & W. A. GODDARD

Orbital optimization in electronic wave functions; equations for quadratic and cubic convergence of general multiconfiguration wave functions

Phys. Rev. A **13**, 1682 (1976)

Although this paper was not widely heralded at the time of publication, it turned out to be the first step in a genuine revolution with respect to the MCSCF method. All conventional SCF methods display linear convergence, since the energy is expanded to first order in the orbital variations. Although Goddard's group had earlier discussed the possibility of quadratic orbital convergence,[a] this elusive goal was achieved for the first time in the present paper.[b] Considering that Yaffe was only an undergraduate at the time, his contribution was all the more remarkable. Yaffe and Goddard showed that by expanding a unitary transformation in terms of completely independent rotation angles, it is possible to derive variational equations correct to any order for MCSCF wavefunctions. The specific MCSCF example presented was for the lowest singlet state of FeO_2, with the argon core of the iron atom replaced by a pseudopotential. Three electron pairs were correlated via the GVB method and the iterative procedure begun from a partially converged calculation. The linear, quadratic, and cubic-convergent methods required ~10, 4, and 3 iterations, respectively, to converge the energy to 10^{-6} hartree.

[a] W. J. Hunt, W. A. Goddard, and T. H. Dunning, The incorporation of quadratic convergence into open-shell self-consistent field equations, *Chem. Phys. Lett.* **6**, 147 (1970).

[b] Note that this formulation does not include the coupling elements between orbitals and CI coefficients. Thus the first truly quadratically convergent methods were those of Dalgaard, Jorgensen, and Yeager (1979).

127

C. F. BUNGE

Accurate determination of the total electronic energy of the Be ground state

Phys. Rev. A **14**, 1965 (1976)

The reader will recall that in 1968 Bunge reported a 180 term CI wavefunction for beryllium accounting for 97.1 per cent of the correlation energy. Eight years later he could report sufficient progress in CI methodology that it was possible to determine the correlation energy significantly more accurately from theory than from experiment. The final estimate was $E_c(\mathrm{Be}) = -0.094\,305 \pm 0.000\,025$ hartree, and no less than 99.6 per cent of this was recovered in Bunge's 650 term CI. A large (10s 9p 8d 7f 5g 3h 1i) and carefully optimized Slater basis set was used in conjunction with sophisticated natural orbital techniques. Since Be is the only system of more than three electrons for which relatively complete Hylleraas-type (explicit incorporation of inter-electronic coordinates in Ψ) wavefunctions are available, comparison with the work of Sims and Hagstrom[a] was of great interest. The latter authors' 107-term wavefunction (their most complete) yielded 99.2 per cent of the correlation energy. Thus the four-electron system is the largest for which the Hylleraas method presently remains competitive with CI.

[a] J. S. Sims and S. A. Hagstrom, Combined CI-HY studies of atomic states. II. Compact wave functions for the Be ground state, *Int. J. Quantum Chem.* 9, 149 (1975).

128

I. SHAVITT

Graph-theoretical concepts for the unitary group approach to the many-electron correlation problem

Int. J. Quantum Chem. Symp. **11**, 131 (1977)

The significance of Paldus's work on the unitary group approach was certainly not quickly appreciated by most chemical theorists. However, Shavitt had previously collaborated with Paldus and Cizek on the coupled cluster method and rather early saw the promise of the UGA. Here Shavitt developed a compact representation of the Gelfand states (configurations in ordinary CI parlance) using a 'distinct row table' or DRT. More important, he introduced a *graphical* representation, which is a pictorial analogue of the DRT and its Gelfand states. The determination of hamiltonian matrix elements was then discussed in terms of graph-theoretical concepts. In a second important paper,[a] one year later Shavitt was able to report much new progress in the formulation of what is now known as the graphical unitary group approach (GUGA). Although Shavitt had not implemented the new method at this stage of its development, he suggested that for large correlated wavefunctions, GUGA would be very effective in the context of a direct CI, following Roos. This prediction has of course been abundantly verified, as Shavitt's graphical representation planted the seeds of a genuine revolution in the CI formalism.

[a] I. Shavitt, Matrix element evaluation in the unitary group approach to the electron correlation problem, *Int. J. Quantum Chem. Symp.* **12**, 5 (1978).

129

W. VON NIESSEN,
G. H. F. DIERCKSEN, &
L. S. CEDERBAUM

On the accuracy of ionization potentials calculated by Green's functions

J. Chem. Phys. **67**, 4124 (1977)

Following McKoy's pioneering work on the equations-of-motion method, a large number of applications of related many-body methods began to appear. Certainly among the most successful were the Green's functions studies of Cederbaum[a] and his co-workers, particularly von Niessen and Diercksen. As the formalism was extended to higher orders of perturbation theory, systematic studies began to establish the reliability to be expected from such methods. The present paper is especially illuminating in this regard since the theory had by 1977 matured, with the development of Cederbaum's diagrammatic third-order approach, augmented by a renormalization procedure to include some higher order contributions. In addition, large basis sets were used here, precluding to a significant degree the fortuitous cancellation of errors. Thus we see, for example, in the limit of a very large basis set that the N_2 Green's function ionization potentials are 15.52 ($3\sigma_g$), 16.83 ($1\pi_u$) and 18.98 eV ($2\sigma_u$), in close agreement with experiment, 15.60, 16.98, and 18.78 eV, respectively. Important related work on ionization potentials using ordinary third-order Rayleigh–Schrödinger perturbation theory should also be mentioned in this context.[b]

[a] Perhaps the first application of the Green's function method to molecular systems was that of L. S. Cederbaum, G. Hohlneicher, and S. D. Peyerimhoff, Calculation of the vertical ionization potentials of formaldehyde by means of perturbation theory, *Chem. Phys. Lett.* **11**, 421 (1971).

[b] D. P. Chong, F. G. Herring, and D. McWilliams, Perturbation corrections to Koopmans' theorem. II. A study of basis set variaton, *J. Chem. Phys.* **61**, 958 (1974).

130

R. P. SAXON & B. LIU

Ab initio configuration interaction study of the valence states of O_2

J. Chem. Phys. **67**, 5432 (1977)

The dissociation of diatomic molecules to their proper separated atom limits has long been recognized as a challenging problem for theory. For example, Schaefer's 1971 study of the O_2 ground state showed that the quadruple excitation $3\sigma_g^2\ 1\pi_u^2 \rightarrow 3\sigma_u^2\ 1\pi_g^2$ is required for dissociation to two ground state Hartree–Fock oxygen atoms. Such situations are poorly described not only by the Hartree–Fock method, but to a less extreme degree by *any* perturbation or coupled cluster scheme based on a single reference configuration. The solution to such problems lies with the more general CI schemes, an excellent example of which is provided by Saxon and Liu's study of 62 low-lying electronic states of the oxygen molecule.[a] In this work a large Slater basis was used, in conjunction with a single set of molecular orbitals (determined from a 9 configuration MCSCF ground state treatment) for each internuclear separation. First-order wavefunctions, including up to 5444 configurations, not only guarantee proper dissociation to the six possible atomic limits but also gave dissociation energies and spectroscopic constants in good agreement with experiment.

[a] A rudimentary study of the same system was reported earlier by H. F. Schaefer and F. E. Harris, *Ab initio* calculations on 62 low-lying states of the O_2 molecule, *J. Chem. Phys.* **48**, 4946 (1968).

131

J. E. GREADY, G. B. BACSKAY, & N. S. HUSH

Finite-field method calculations. IV. Higher-order moments, dipole moment gradients, polarisability gradients, and field-induced shifts in molecular properties: application to N_2, CO, CN^-, HCN, and HNC

Chem. Phys. **31**, 467 (1978)

The dipole moment derivatives of HCN posed a serious challenge to theory for most of the 1970s and suggested to many observers that such predictions might in general be unreliable. The 1952 infrared intensity studies of Hyde and Hornig[a] showed that μ'(H–C) and μ'(C≡N) had opposite signs, but theoretical studies both semi-empirical and *ab initio* found the dipole moment to increase with the elongation of *both* bonds. Since Brun and Person's predictions[b] of μ' were based on McLean and Yoshimine's near Hartree–Fock wavefunctions, there were serious doubts raised that either the data or the interpretation of the 1952 experiment might be in error. However, a new analysis using some new experimental data found μ'(H–C) = ±0.21 and μ'(C≡N) = ±0.059 atomic units. The problem was finally resolved by Gready, Bacskay, and Hush, who demonstrated that electron correlation effects change the sign of μ'(C≡N). An independent theoretical study by Liu and co-workers[c] confirmed the Australian result.

[a] G. E. Hyde and D. F. Hornig, The measurement of bond moments and derivatives in HCN and DCN from infrared intensities, *J. Chem. Phys.* **20**, 647 (1952).
[b] R. E. Bruns and W. B. Person, Calculated dipole-moment functions and infrared intensities of HCN and N_2O, *J. Chem. Phys.* **53**, 1413 (1970).
[c] B. Liu, K. M. Sando, C. S. North, H. B. Friedrich, and D. M. Chipman, Theoretical dipole moment derivatives and force constants for HCN, *J. Chem. Phys.* **69**, 1425 (1978).

132

J. A. POPLE, R. KRISHNAN, H. B. SCHLEGEL, & J. S. BINKLEY

Electron correlation theories and their application to the study of simple reaction potential surfaces

Int. J. Quantum Chem. **14**, 545 (1978)

A distinctive feature of Pople's work on the correlation problem during this period was the development of a working perturbation theory without diagrams. This of course was a great relief to the many chemists who considered diagrams to be at best a distraction. Also the notion of including certain contributions to high order while neglecting others appearing in lower orders of perturbation theory seemed questionable. Pople's systematic treatment followed the outline provided by Møller and Plesset, who derived a simple, explicit expression for the second-order energy. A computationally practical third-order energy expression was first explicitly obtained by Bartlett and Silver,[a] albeit using diagrammatic techniques. In fourth order, single, triple, and quadruple excitations appear for the first time, and the treatment of single (as well as double) excitations is straightforward. Quadruple excitations are much more difficult to deal with in fourth order, but Krishnan and Pople[b] used non-diagrammatic manipulations to provide working expressions for quadruples in 1978. This paper and the following by Bartlett and Purvis present some of the earliest MP4 (SDQ)[c] results for molecules.

[a] R. J. Bartlett and D. M. Silver, Many-body perturbation theory applied to electron pair correlation energies. I. Closed-shell first-row diatomic hydrides, *J. Chem. Phys.* **62**, 3258 (1975).

[b] R. Krishnan and J. A. Pople, Approximate fourth-order peturbation theory of the electron correlation energy, *Int. J. Quantum Chem.* **14**, 91 (1978).

[c] Using Pople's standard nomenclature, MP4 (SDQ) describes the use of fourth-order Møller–Plesset perturbation theory in the space of single, double, and quadruple excitations.

133

R. J. BARTLETT & G. D. PURVIS

Many-body perturbation theory, coupled-pair many-electron theory, and the importance of quadruple excitations for the correlation problem

Int. J. Quantum Chem. **14**, 561 (1978)

Although the coupled cluster method was formulated for electronic problems by Cizek in 1966, very little development of the method took place for more than a decade. This paper and the previous one by Pople's group provided the first general implementations of the coupled cluster method with double substitutions (CCD). It may be noted that Pople's interest in the CCD method was spurred by the clear exposition given in Hurley's book.[a] From a practical perspective the CCD wavefunction and energy are obtained via an iterative procedure, which requires about twice the effort needed for the comparable CI including all double excitations (CID). By comparison with perturbation theory, the CCD expansion is correct to third order, but at fourth order includes correct contributions only from double and quadruple excitations. Single and triple excitations are of course ignored in the CCD method. The actual calculations showed that the CCD energy was always lower than the comparable CI result, but quite comparable to the MP4 (DQ) energy. Since MP4 (DQ) may be viewed as a fourth-order truncation of CCD, the higher orders of perturbation theory would appear relatively unimportant.

[a] See pages 204–21 of A. C. Hurley, *Electron correlation in small molecules* (Academic, New York, 1976).

134

P. SIEGBAHN & B. LIU

An accurate three-dimensional potential energy surface for H_3

J. Chem. Phys. **68**, 2457 (1978)

The H + H_2 exchange reaction has since the work of Hirschfelder and Eyring in the 1930s been considered the prototype A + BC reaction. Great strides were taken in the *ab initio* characterization of this energy surface in 1956 and 1968 with the paper of Boys, Shavitt, and Karplus. For linear H_3, Liu was able to report in 1973 a definitive surface,[a] estimated to lie between 0.2 and 0.8 kcal above the exact non-relativistic limit. However, Liu's CI wavefunctions were constructed from a Slater basis set and the extension to non-linear geometries was not straightforward. Therefore a contracted gaussian basis set H(9s 3p 1d/4s 3p 1d) was used here to span the entire energy hypersurface. Full CI was carried out using a specially designed direct CI method developed by Siegbahn. 156 different geometries were considered and an accompanying paper[b] reported a rather precise analytic fit to the Siegbahn–Liu H_3 surface. This surface is surely accurate to within better than one kcal in the low energy region and has been used in a range of dynamical studies of the H_3 system.

[a] B. Liu, *Ab initio* potential energy surface for linear H_3, *J. Chem. Phys.* 58, 1925 (1973).

[b] D. G. Truhlar and C. J. Horowitz, Functional representation of Liu and Siegbahn's accurate *ab initio* potential energy calculations for H + H_2, *J. Chem. Phys.* 68, 2466 (1978).

135

R. J. BUENKER, S. D. PEYERIMHOFF, & W. BUTSCHER

Applicability of the multi-reference double-excitation CI (MRD–CI) method to the calculation of electronic wavefunctions and comparison with related techniques

Mol. Phys. **35**, 771 (1978)

Buenker and Peyerimhoff developed their MRD–CI method over a period of several years,[a] but this paper may be taken as their definitive statement. The basic approach is to consider all double excitations relative to a multi-configuration reference function, the latter including typically 10 configurations. Each double excitation is tested by the magnitude of its hamiltonian matrix element with the reference function, and those configurations surpassing a certain threshold[b] are included in the final variational treatment. The wavefunctions obtained in this way provide a large fraction of the correlation energy with relatively small numbers of configurations. By simultaneously increasing the size of the reference function and decreasing the energy-threshold one will eventually approach the full CI wavefunction. In practice, Buenker and Peyerimhoff use extrapolation procedures based on extensive experience to estimate the approach to the full CI energy. Employed in this way the MRD–CI method has been eminently successful in treating a range of chemical problems.

[a] For earlier work, see R. J. Buenker and S. D. Peyerimhoff, Individualized configuration selection in CI calculations with subsequent energy extrapolation, *Theoret. Chim. Acta* **35**, 33 (1974).

[b] An attractive alternative is the *cumulative* selection procedure. There the sum of the perturbation contributions due to the deleted configurations is constrained to be constant over the potential energy surface being studied. See R. C. Raffenetti, K. Hsu, and I. Shavitt, Selection of terms for a CI wavefunction to preserve potential surface features, *Theoret. Chim. Acta* **45**, 33 (1977).

136

J. A. POPLE, R. KRISHNAN, H. B. SCHLEGEL, & J. S. BINKLEY

Derivative studies in Hartree–Fock and Møller–Plesset theories

Int. J. Quantum Chem. Symp. **13**, 225 (1979)

The first practical[a] *ab initio* formulation of Hartree–Fock analytical second derivatives was presented here by the Pople group. Also important (but less so) was the introduction of energy first derivatives at the simplest correlated level of theory, second-order Møller–Plesset perturbation theory. To obtain SCF energy second derivatives, one must first evaluate the second derivatives of all one- and two-electron integrals, and this was achieved using the method of Rys polynomials (Dupuis, Rys, and King 1976). Secondly, one must obtain the first-order changes in the SCF molecular orbitals, via the coupled-perturbed Hartree–Fock (CPHF) equations of Gerratt and Mills (1968). Perhaps the most important theoretical development in the paper of Pople *et al.*, is the fomulation of an efficient iterative method for the solution of these CPHF equations. The test case chosen was the ethylene ground state with a 6-31*G** basis set (38 contracted gaussians), and both the MP2 first derivatives and SCF second derivatives were obtained much more rapidly than would have been the case using numerical finite difference methods.

[a] An earlier, less practical implementation was that of K. Thomsen and P. Swanstrom, Calculation of molecular one-electron properties using coupled Hartree–Fock methods. I. Computational scheme, *Mol. Phys.* **26**, 735 (1973).

137

B. R. BROOKS & H. F. SCHAEFER

The graphical unitary group approach to the electron correlation problem. Methods and preliminary applications

J. Chem. Phys. **70** 5092 (1979)

This paper presented the first general implementation of Shavitt's graphical unitary group approach. The authors concluded that the power of the concepts developed by Paldus and Shavitt lies with the tremendous amount of insight provided into the structure of the hamiltonian matrix. The specific formulation of GUGA reported by Brooks later came to be called the *loop-driven* method,[a] since it is structured with respect to common loops on the Shavitt graph, rather than with respect to integrals or configurations. While concurring with Shavitt's prediction that GUGA is very suitable for a direct CI approach, Brooks showed that for modest CI (up to perhaps 25 000 configurations) the implementation within the traditional 'formula tape' framework yields an exceptionally efficient method. In addition to restructuring the hamiltonian in terms of loop types, this paper provides the solution to the upper walk (walk = configuration) problem, by reversing the lexical ordering introduced by Shavitt. A final important innovation was the incorporation of the multireference interacting space (see A. Bunge 1970) into the framework of GUGA.

[a] B. R. Brooks, W. D. Laidig, P. Saxe, N. C. Handy, and H. F. Schaefer, The loop-driven graphical unitary group approach: a powerful method for the variational description of electron correlation, *Physica Scripta* **21**, 312 (1980).

138

P. E. M. SIEGBAHN

Generalizations of the direct CI method based on the graphical unitary group approach. I. Single replacements from a complete CI root function of any spin, first order wave functions

J. Chem. Phys. **70**, 5391 (1979)

Brooks' development of GUGA was followed very shortly by Siegbhan's method, which in several ways is more typical of the current state-of-the-art. Roos, with Siegbahn, had been developing the direct CI method for the previous seven years, and the latter seized upon GUGA as the device to effect a generalization of the direct CI. Although Siegbahn's initial report was restricted to first-order wavefunctions (Schaefer and Harris (1968)), a second paper quickly appeared[a] extending the method greatly. In retrospect, the most important element of Siegbahn's landmark paper is the drastic reduction of the 'coupling coefficients', which show how each integral appears in the iterative matrix diagonalization scheme. If the internal space is defined as those orbitals occupied in one of the reference configurations, then this space will typically be only a fraction of the complete orbital basis. By restricting the wavefunction to single and double excitations relative to such a multi-configuration reference function, the coupling coefficients involving the external orbitals are found to have a simple structure. Thus all coupling coefficients may be reduced to expressions involving only the small internal space. The importance of this discovery is attested to by the fact that it plays a central role in the current methods of both the Ohio State[b] and Berkeley[c] groups.

[a] P. E. M. Siegbahn, Generalizations of the direct CI method based on the graphical unitary group approach. II. Single and double replacements from any set of reference configurations, *J. Chem. Phys.* **72**, 1647 (1980).

[b] H. Lischka, R. Shepard, F. B. Brown, and I. Shavitt, New implementation of the graphical unitary group approach for multireference direct configuration interaction calculations, *Int. J. Quantum Chem. Symp.* **15**, 91 (1981).

[c] P. Saxe, D. J. Fox, H. F. Schaefer, and N. C. Handy, The shape-driven graphical unitary group approach to the electron correlation problem. Application to the ethylene molecule, *J. Chem. Phys.* **77**, 5584 (1982).

139

B. O. ROOS, P. R. TAYLOR, & P. E. M. SIEGBAHN

A complete active space SCF method (CASSCF) using a density matrix formulated super-CI approach

Chem. Phys. **48**, 157 (1980)

Although this paper appeared in 1980, it is to be emphasized that it represents a complete report, and that Roos's work in this regard was well known (through lectures) for some time prior to the appearance of this paper. This is an important point, because the significance of the CASSCF method was somewhat overshadowed by the later appearance of several quadratically convergent MCSCF formalisms. It should also be clearly stated that Ruedenberg's group was developing a closely-related method during the same general time frame.[a] The essence of the Roos–Taylor–Siegbahn approach is to define a small set of valence orbitals within which the most important correlation effects (for example, those required to obtain proper molecule dissociation) are expected to occur. Within this small space of *active* molecular orbitals a full CI is constructed and the MCSCF procedure used to minimize the energy both with respect to orbitals and CI coefficients. Most of the early large (>100 configurations) MCSCF wavefunctions were determined by the CASSCF method. The only drawback to the method is that the size of the full CI goes up very rapidly, even for a relatively small number of valence electrons.

[a] For a high-level, recent discussion see K. Ruedenberg, M. W. Schmidt, M. M. Gilbert, and S. T. Elbert, Are atoms intrinsic to molecular electronic wavefunctions? I. The FORS model, *Chem. Phys.* **71**, 41 (1982).

140

B. LIU & A. D. McLEAN

Ab initio potential curve for $Be_2(^1\Sigma_g^+)$ from the interacting correlated fragments method

J. Chem. Phys. **72**, 3418 (1980)

The description of weakly interacting chemical systems has long been a major challenge to theory, and the paper by Liu and McLean made significant progress in this respect. The Be_2 system was a particularly intriguing one since previous theoretical studies gave potential minima from ~2.5-5.0 Å, an enormous range. The approach taken by Liu and McLean was to describe each separate Be atom with the TCSCF wave-function

$$\Psi = c_1\, 1s^2 2s^2 + c_2\, 1s^2 2p^2,$$

which adds to the Hartree-Fock configuration the important near-degeneracy effect $2s^2 \rightarrow 2p^2$. This requires a sizeable molecular MCSCF treatment with four core electrons ($1\sigma_g^2\, 1\sigma_u^2$) followed by a complete set of $D_{\infty h}$ symmetry-restricted configurations with the four valence electrons in the valence orbitals $2\sigma_g$, $2\sigma_u$, $3\sigma_g$, $3\sigma_u$, $1\pi_u$, and $1\pi_g$. Then a transformation to localized orbitals is carried out and accompanied by a large CI including inter-atomic correlation effects. In this sense the method may be viewed as a generalization of the earlier discussed 1970 studies of He_2, but with the separated atoms described in a MCSCF rather than SCF manner. The Liu-McLean method thus allows the inclusion of the most important intra-atomic correlation effects as well. Still more exhaustive recent studies[a] of Be_2 have confirmed the predictions (r_e = 2.49 Å, D_e = 0.1 eV) of Liu and McLean in detail.

[a] B. H. Lengsfield, A. D. McLean, M. Yoshimine, and B. Liu, The binding energy of the ground state of Be_2, *J. Chem. Phys.* **79**, 1891 (1983).

141

R. KRISHNAN, M. J. FRISCH, & J. A. POPLE

Contribution of triple substitutions to the electron correlation energy in fourth order perturbation theory

J. Chem. Phys. 72, 4244 (1980)

Here Krishnan, Frisch, and Pople completed the formulation and implementation of full fourth-order peturbation theory without diagrams. Only later were diagrammatic MBPT techniques able to match this important achievement.[a] In previous work the most difficult terms to treat were the fourth-order quadruple excitations, for which effort proportional to n^6 (n = number of orbitals) is required. It may be noted that the CI including all single and double excitations likewise involves $O(n^6)$ operations. The triple substitutions, however are inherently more diffiult to treat and even in the earliest perturbation order in which they appear (fourth), their theoretical treatment requires $O(n^7)$ operations. Although some workers had assumed triple excitations to be of insignificant importance, Pople's work[b] showed this not to be the case. Most strikingly, for the N_2 molecule, with a roughly DZ + P basis set, the fourth-order perturbation contributions were: singles −0.0039, doubles plus quadruples −0.0040, and triples −0.0133 hartree. The fact that the triples are more than three times as important as doubles plus quadruples is evident. Moreover for N_2 the triples served to increase the predicted dissociation energy by a staggering 7.8 kcal.

[a] M. F. Guest and S. Wilson, Triple and quadruple excitations and the valence correlation energies of small molecules, *Chem. Phys. Lett.* 72, 49 (1980).

[b] See also M. J. Frisch, R. Krishnan, and J. A. Pople, A systematic study of the effect of triple substitutions on the electron correlation energy of small molecules, *Chem. Phys. Lett.* 75, 66 (1980).

142

B. R. BROOKS, W. D. LAIDIG,
P. SAXE, J. D. GODDARD,
Y. YAMAGUCHI &
H. F. SCHAEFER

Analytic gradients from correlated wave functions via the two-particle density matrix and the unitary group approach

J. Chem. Phys. 72, 4652 (1980)

The earliest development of the analytic gradient method for CI wavefunctions is reported in this and the following paper. The Brooks formulation of the CI gradient leaned heavily upon the loop-driven graphical unitary group approach described earlier. Specifically, it was found that with the GUGA perspective on matrix element determination, it was possible to obtain the two-particle density matrix for large correlated wavefunctions very rapidly. Previously it had been assumed that the two-particle density matrix required nearly the same amount of computation as needed to determine the CI wavefunction at the outset. A subsequent back-transformation of the density matrices then provided a transparent expression for the analytic forces in terms of the derivative integrals, Lagrangian matrix, and first-order changes in the SCF molecular orbitals. The method was restricted to CI wavefunctions constructed from closed-shell SCF orbitals, since the coupled-perturbed Hartree–Fock method developed was limited to that special case. A full report of this work appeared shortly.[a]

[a] B. R. Brooks, W. D. Laidig, P. Saxe, J. D. Goddard, and H. F. Schaefer, New directions for the loop-driven graphical unitary group approach: analytic gradients and an MCSCF procedure, pages 158–76 of Volume 22, *Lecture notes in chemistry*, editor J. Hinze (Springer-Verlag, Berlin, 1981).

143

R. KRISHNAN, H. B. SCHLEGEL, & J. A. POPLE

Derivative studies in configuration-interaction theory

J. Chem. Phys. **72**, 4654 (1980)

Pople's formulation of CI energy first derivatives is rather different from the previous treatment. The CI wavefunction is specifically restricted to all single and double spin-orbital excitations relative to a single-determinant SCF wavefunction. It should be emphasized, of course, that this is precisely the most frequently used type of CI for studies of chemical systems. Within this definite theoretical framework, Krishnan, Schlegel, and Pople were able to derive an explicit analytic expression for the CI forces in terms of the integral derivatives, the CI coefficients, and the first-order changes in the SCF molecular orbitals. Both this paper and the previous CI derivative scheme were able to report impressive computation times. While the numerical determination of forces via finite difference requires $6N$ evaluations of the total energy (N = number of atoms in the molecule), the analytic CI derivative methods required less additional time than the original CI. A summary of the Carnegie-Mellon experience with derivative methods was later given in a review by Schlegel.[a]

[a] H. B. Schlegel, *Ab initio* energy derivatives calculated analytically, pages 129–59 of *Computational theoretical organic chemistry*, edited by I. G. Csizmadia and R. Daudel (D. Reidel, Dordrecht, Holland, 1981).

144

B. H. LENGSFIELD

General second order MCSCF theory: a density matrix directed algorithm

J. Chem. Phys. **73**, 382 (1980)

Certainly the most important development in MCSCF theory in the past decade has been the introduction of several functioning formalisms for the achievement of quadratic convergence. This is tremendously important, since each MCSCF iteration includes a CI calculation, and the attendant four-index transformation of two-electron integrals is of course very time-consuming. Thus the reduction of the MCSCF procedure to the minimum possible number of iterations is a primary goal for the method. In this reviewer's mind, the first general (i.e., arbitrary configuration list) second-order MCSCF method capable of treating more than a few configurations was documented in this paper by Lengsfield. A virtue of his 'density-matrix directed' approach is that it avoids the horrendous complexity associated with attempting to explicitly construct the MCSCF energy expression for each new set of configurations. As a test case Lengsfield chose the notoriously difficult[a] BeO $^1\Sigma^+$ ground state and was able to report worst-case energy convergence to 10^{-6} hartrees in eight iterations (for an 81 configuration wavefunction). The time required to generate the hessian of the energy expression (required for MCSCF optimization) was significantly less than the conventional CI work required during each MCSCF iteration.

Unfortunately, a passage of time must occur before an impartial appraisal of the development of MCSCF second-order methods is possible. In this context we must cite several other papers which may come to be valued as highly as Lengsfield's. First is Dalgaard and Jorgensen's 1978 paper,[b] which gives a rather complete derivation of the 'exponential operator method'. This method was applied simultaneously by Dalgaard[c] and by Yeager and Jorgensen[d] to small MCSCF calculations. Shortly thereafter Shepard and Simons[e] independently presented a third quadratically convergent MCSCF method and Werner and Meyer[f] a fourth working formalism. In quick succession Lengsfield and Liu[g] extended the second-order MCSCF methods to much larger cases by iteratively solving the linear Newton–Raphson equations. A detailed comparison of the different iterative methods is given in the recent paper of Shepard, Shavitt, and Simons.[h] We commend all of these papers to the reader.

[a] C. W. Bauschlicher and D. R. Yarkony, MCSCF wavefunctions for excited states of polar molecules: application to BeO, *J. Chem. Phys.* **72**, 1138 (1980).

[b] E. Dalgaard and P. Jorgensen, Optimization of orbitals for multiconfigurational reference states, *J. Chem. Phys.* **69**, 3833 (1978).

[c] E. Dalgaard, A quadratically convergent reference state optimization procedure, *Chem. Phys. Lett.* **65**, 559 (1979).

[d] D. L. Yeager and P. Jorgensen, Convergency studies of second and approxi-

mate second order multiconfigurational Hartree–Fock procedures, *J. Chem. Phys.* **71**, 755 (1979).

[e] R. Shepard and J. Simons, Multiconfigurational wavefunction optimization using the unitary group method, *Int. J. Quantum Chem. Symp.* **14**, 211 (1980).

[f] H. -J. Werner and W. Meyer, A quadratically convergent MCSCF method with simultaneous optimization of orbitals and CI coefficients, *J. Chem. Phys.* **73**, 2342 (1980).

[g] B. H. Lengsfield and B. Liu, A second order MCSCF method for large CI expansions, *J. Chem. Phys.* **75**, 478 (1981).

[h] R. Shepard, I. Shavitt, and J. Simons, Comparison of the convergence characteristics of some iterative wave function optimization methods, *J. Chem. Phys.* **76**, 543 (1982).

145

Y. OSAMURA, Y. YAMAGUCHI, P. SAXE, M. A. VINCENT, J. F. GAW, & H. F. SCHAEFER

Unified theoretical treatment of analytic first and second energy derivatives in open-shell Hartree–Fock theory

Chem. Phys. **72**, 131 (1982)

Pople's analytic second derivative method for single determinant UHF wavefunctions was presented at the March, 1979 Sanibel Symposium at Palm Coast, Florida. The fact that no other theoretical group developed a second derivative method until more than three years thereafter indicated (a) the inherent difficulty of putting together any such new method and (b) the inherent difficulty of formulating and solving the CPHF equations for cases more complicated than the UHF case of Pople. As we will note shortly Osamura and Yamaguchi were successful in the resolution of problem (b). It is thus not surprising that within one year, these workers were able to exploit the same CPHF method (used earlier for CI *first* derivatives) for the determination of analytic *second* derivatives for nearly arbitary RHF wavefunctions, including the notorious open-shell singlet case. The authors were able to show the greatly improved efficiency of the analytic SCF second derivative method relative to the traditional alternative of obtaining force constants as finite differences of analytic gradients. Moreover, they demonstrated that for some electronic states the analytic second derivative technique avoids the variational collapse accompanying the traditional finite difference SCF gradient approach.

146

G. D. PURVIS & R. J. BARTLETT

A full coupled-cluster singles and doubles model: the inclusion of disconnected triples

J. Chem. Phys. **76**, 1910 (1982)

We have seen that the coupled cluster doubles (CCD) model was developed in working form by the groups of Pople and of Bartlett in 1978. In one sense, however, the CCD method is not entirely hierarchical. Although single excitations do not appear in perturbation theory until fourth order, they are in another perspective more closely related to the Hartree–Fock reference function than are double excitations. Moreover, the famous case of the CO dipole moment showed that single excitations can have a large effect on predicted one-electron properties. Therefore a perhaps more consistent (and obviously more complete) model is the coupled cluster approach with

$$\hat{T} = \hat{T}_1 + \hat{T}_2,$$

wherein both one- and two-body excitation operators are considered. In this way the exponential ansatz includes all single excitations and in addition triple (and higher) excitations due to disconnected products of singles and doubles. Although CCSD is not a variational method, it is size-consistent and grows no more rapidly than the sixth power of the number of basis functions. Purvis and Bartlett presented results for H_2O and BeH_2 and made careful comparison with available full CI results. This writer would not be surprised if the CCSD method becomes as widely used during the next decade as the CISD method has been during the past.

147

Y. OSAMURA, Y. YAMAGUCHI, & H. F. SCHAEFER

Generalization of analytic configuration interaction (CI) gradient techniques for potential energy hypersurfaces, including a solution to the coupled perturbed Hartree–Fock equations for multiconfiguration SCF molecular wave functions

J. Chem. Phys. **77**, 383 (1982)

The earlier-discussed analytic CI gradient methods were restricted with respect to the type of correlated wavefunction that could be used. In Brooks's formulation any configuration set allowed within their unitary group approach could be treated, but the molecular orbitals employed had to be closed-shell restricted Hartree–Fock (RHF) orbitals following Roothaan (1951). In Pople's approach unrestricted Hartree–Fock orbitals were used in conjunction with CI involving all single and double spin-orbital excitations relative to the single determinant reference function. In this paper Osamura and Yamaguchi generalized the analytic CI gradient method to handle essentially any CI constructed from nearly arbitrary RHF orbitals. The most difficult case thus treated was the open-shell singlet, although the formalism developed allows the treatment of any state with energy expression dependent only on the integrals h_i, J_{ij}, and K_{ij}. The method was tested by comparison with finite difference results for the first excited singlet state of methylene, and the vibrational frequencies of this species subsequently predicted. A further development, of a more mathematical nature, was the formulation of the CPHF equations for MCSCF wavefunctions. The first application of this new theory has recently appeared.[a]

[a] Y. Yamaguchi, Y. Osamura, G. Fitzgerald, and H. F. Schaefer, Analytic force constants for post-Hartree-Fock wavefunctions: the simplest case, *J. Chem. Phys.* 78, 1607 (1983).

148

J. ALMLÖF, K. FAEGRI, & K. KORSELL

Principles for a direct SCF approach to LCAO–MO *ab initio* calculations

J. Comput. Chem. **3**, 385 (1982)

As one proceeds to near Hartree-Fock studies of larger and larger molecules, the number of two-electron integrals to be stored for the SCF iterative process becomes staggering. Ultimately, and especially for systems of low symmetry (not uncommon for large molecules), the number of electron repulsion integrals may become too large to be handled by any reasonably responsive input/output device (disk, tape, etc.). Put more succinctly, the effort required to retrieve such a large list of integrals during the SCF iterative procedure becomes greater than that required to compute the integrals in the first place. This in essence was the motivation behind Almlöf's development of the 'direct SCF' method, in which integrals are evaluated as required during each SCF iteration. These workers showed that the method is far more attractive than might at first appear, since the density matrix is known at each iteration and may be used to eliminate large numbers of integrals. This is because each Fock matrix contribution is the *product* of a density matrix element and a two-electron integral. The effectiveness of the new method was proven by an SCF treatment of decamethyl ferrocene using a triple zeta basis set (501 contracted gaussian functions).

149

V. R. SAUNDERS & J. H. VAN LENTHE

The direct CI method. A detailed analysis

Mol. Phys. **48**, 923 (1983)

The world's first production supercomputer, the Cray-1, was delivered to the Los Alamos Scientific Laboratory (New Mexico) on April Fool's Day, 1976. For a considerable period of time, it was not obvious to theoretical chemists what was so 'super' about this particular machine. More than two years after its appearance at Los Alamos, a sympathetic report[a] indicated that speeds for scalar operations were 2.3–2.8 times those achieved on the older Control Data 7600, while for vectorizable operations, factors of five could be achieved in some cases. Given the complexity of machine language programming for the Cray-1, the rewards were not appealing. That assessment changed radically a few years after the arrival of the Cray-1S (same machine, more memory) at the SERC Daresbury Laboratory, England. There Saunders and Guest[b] were able to show that most molecular electronic structure algorithms may be revised to allow domination by a series of (vectorizable) matrix multiplications. In this way speed enhancement of factors of five to *twenty-five* over the CDC 7600 were achieved. The paper by Saunders and van Lenthe shows how a truly remarkably efficient CI method was designed specifically for the Cray-1S supercomputer.

[a] F. W. Dorr, The Cray-1 at Los Alamos, *Datamation*, pp. 113–20, October (1978).
[b] V. R. Saunders and M. F. Guest, Applications of the Cray-1 for quantum chemistry calculations, *Comput. Phys. Comm.* **26**, 389 (1982).

Author Index

AUTHOR INDEX

Subject Index

SUBJECT INDEX

SUBJECT INDEX